Peter Suciu is a freelance writer and author. His work has appeared in more than three dozen magazines, newspapers and websites including *Newsweek, Playboy, Reader's Digest, Consumer's Digest, Forbes, The New York Daily News, The New York Post*, FoxNews.com, CNN.com and MSNBC.com.

Peter is also a well respected military writer and for more than a decade has covered military history and militaria. In that time he has written for *Armchair General, Military Heritage, Military Trader and World War II History*.

Since 1981 Peter has been an avid collector of militaria and especially military helmets, of which he has an extensive collection. He has had a lifetime passion for military sun helmets, a much underrated area of military history, and began collecting these helmets from such diverse countries as England to Thailand to Paraguay. This book is a result of Peter's desire to see sun helmets recognised for the important part that they have played in military history.

Dedicated to Louis Suciu, my great uncle and World War II veteran, who bought me my first helmet in October 1981.

MILITARY SUN HELMETS OF THE WORLD

by Peter Suciu
with Stuart Bates

The Naval & Military Press Ltd

Published by

The Naval & Military Press Ltd
Unit 10 Ridgewood Industrial Park,
Uckfield, East Sussex,
TN22 5QE England
Tel: +44 (0) 1825 749494
Fax: +44 (0) 1825 765701
www.naval-military-press.com
www.military-genealogy.com
www.militarymaproom.com

ISBN 978-1-845749-12-5

Cover design by Nick Jenkin

Rear cover:
"Lord William Beresford's encounter with a Zulu across the Umvolosi river"
From the Illustrated London News, 6th September 1879.

Printed and bound in China.

Table of Contents

FOREWORD

I have been a collector of things all of my life. As a young boy I collected baseball, monster, Bit-A-Honey, and other collectable cards. Throw in a couple hundred stamps, coins, seashells, matchbooks, and now militaria and you get an idea of my desire to collect – something. During this period I found that I was always more appreciative of the items with historical significance. Thus out with all but the coin collection. As with many collectors, I had to find ways to acquire the items I wanted for my collection. Usually it meant earning the money from a day job or sell or trade another item from my collection to acquire the new one, thus I found I had to focus on one area of collecting. This happened by sheer chance when I decided to sell my coin collection to acquire a 45-70 Springfield. From there I started collecting items related to the Indian Wars in the United States. I had subscribed to the series of books from Time-Life on the West and found my calling when I saw the photos and color prints depicting the soldiers of the period with their Prussian-style helmets. The first helmet that I acquired came out of a militaria antiques dealer in Portland, Oregon. I was so pleased with it that I started to research everything I could on the helmet. To my dismay only a handful of resources were available. These included monographs from the Arizona Historical Society by Gordon Chappell and Sidney Brinkerhoff. From there I was hooked. I now had my military contact to the Old West. As I became more serious I found that I had purchased many items that were not "correct". Even in the 1980s there were many reproductions and restrikes of uniform parts from the Civil War and Indian Wars, many struck from the original dies. I discovered so much misinformation and so many reproductions that I made it a point to be as educated in what I wanted to collect as I could. From this quest and my partnership with my good friend Don Moore, evolved our book Dress Helmets of the U.S. Army, 1872-1904. Although much of the information included in that book was available from many other sources there was no single source to help the collector and researcher. Peter Suciu has performed a very commendable service for collectors of military sun helmets of the world. Peter brings the same desire and education on this subject that can only be developed by asking the questions, and making the many errors that serious collectors and researchers make. This book is a must for any person interested in military uniforms found in any part of the world.

Mark R. Kasal

"Doctor Livingston, I presume." Those words as much as any other likely sum up sun helmets for most people. The sun helmet, or pith helmet as it is as commonly known, is well-known as the choice of headgear for archeologists, explorers and big game hunters – who if serialized fiction is to be believed wear their "safari garb" to fashionable parties in London's West End.

The helmets have a civilian origin and were gradually adopted by the military, but even to collectors of military headdress the sun helmet is not a combat item. While certainly not part of a dress uniform sun helmets are sort of the "other helmet" to many collectors. However, the sun helmet's history on the battlefield is a long one, and while offering little if any ballistic protection the sun helmets provided adequate protection from the rays above and that was after all its intended purpose.

The helmet was practically born in battle, adopted by the British during the heated Indian Mutiny in 1857 and used to the present day, this form of headdress saw action in numerous colonial conflicts, both World Wars and of course was the definitive headgear used by the Communist forces in the Vietnam War.

It is impossible to think of famous battles such as the British defense of Rorkes Drift, which is used as the setting for the colonial wartime cinematic epic Zulu, without thinking of the white sun helmets. The image in the movie is wrong of course; the British didn't officially wear regimental plates on the front of the helmets. But the helmets are as iconic as the red coats. The sun helmets used by the North Vietnamese Army in Vietnam can also be seen as the object that most defines the enemy, much the way the German steel helmet did for the German Wehrmacht forces in World War II. Here again, badges are seen in the movies and even those in most collections – yet in the field these were probably worn without any form of badge, in part to hide that these were really members of the NVA, and not simple "volunteers" fighting for unification of a nation in civil war.

These are just two examples of how the sun helmet played a role in many conflicts in the past 150 years. Sun helmets were used by all of the major colonial powers in Europe, were adopted by the United States, Canada, many South American nations and several Asian nations. Since the 1870s the sun helmet has served in some military capacity around the world.

The helmet was in wide civilian use, and of course was used by explorers such as Stanley and Livingston. Today it is a standard headdress used by police in warm weather regions, as well as those people going on safari. Sun helmets even remain a popular form of head cover for the United States Postal Service, with the modern plastic versions providing protection from sun and rain alike.

This book looks at the origins, history and evolution of the military sun helmet. This is by no means a definitive or final word on the subject. While great efforts have been made to research the development of helmets of every nation, obviously this was not always possible. New information will certainly come to light and new facts will be uncovered. And while much of what has been written on sun helmets remains scattered, it is the hope of the author to provide a single and concise collector's guide to a type of headgear that deserves a bit more respect.

Peter Suciu
November 2008

Acknowledgements

I would like to thank the following friends for their inspiration and guidance in helping make this book possible: Chris Armold, Zachary Bathon, Marc Barrie, Tom Buck, A. Gustaf Bryngelson, Robert Clawson, Mark Dillenbeck, Warren Everett, David Ford, Keith Gill, Mark Giroux, John Adams-Graf, Robert Hinesley, Ron Kidd, Eduardo Antonio Krapovickas, Clive Law, Gabriel Llado, Frederic Mahout, Paolo Marzetti, Don Moore, Jerry Morelock, Larry Munnikhuysen, Dave Powers, David Read, Joe Robinson, Jeff Shrader, Jean-Paul Soulier, Victor Taboika, Alec Tulkoff and Steve Wilson.

Thank you to Bill Combs, Chris Dale, Jareth Holub, Mark R. Kasal, Chris Mills, Ken Niewiarowcz and Doug Zinn for providing their insight and knowledge on a subject that has only sparsely been written about. Without the insight from these gentlemen this book could not have been possible.

I would like to offer tremendous thanks to Dave Gerardi and Nick Jenkin, each of whom lent their time and talents in assisting with design aspects of this book as well as their photographic expertise. An additional debt must be acknowledged to Nick Jenkin whose undoubted skill and patience in editing every photo in this book has added markedly to its quality.

I also offer a special thank you to my very good friend and colleague Stuart Bates. Without him this book would not exist. In addition to his knowledge of the subject, Stuart provided the design and layout, proofing, research assistance, insight and personal support through all the difficult times. Thank you Stuart.

Finally thanks to my wife Enid for allowing me to drag her through endless military museums and strange militaria shops around the world. Thank you Enid for sticking with me and allowing me to fill our home with so many helmets!

Chapter 1

British and Commonwealth Helmets

Often dubbed the "pith helmet" or "sun helmet", this style of headgear is one of the most evocative pieces of militaria from the Victorian Age. The British Foreign Service Helmet, which was used for the last half of the 19th century, remains today as iconic to the era as the German steel helmet is to World War II. It is a piece that also has eluded any significant study. However, much of the origins of the helmet are still somewhat shrouded in mystery. Yet, it is hard to think of those British soldiers in red coats facing off against the forces of the Mahdi, or standing tall against the Zulu without their Foreign Service Helmets.

Indian Origins

Throughout the first half of the 19th century the British Army continued to wear uniforms that were only slightly changed from the Napoleonic Wars. This included the use of the shako, which remained popular throughout the armies of Europe. In the course of 60 years the British used several different versions of shakos and these continued to be worn in climates where it was anything but appropriate.

By the early 1840s the first sun helmets appeared in India, which was administered by the Honourable East India Company (HEIC). These helmets were locally made, with the materials typically being sola pith – which of course is the origin of the name "pith helmet" – from the flowering plant *Aeschynomene asper* or other species of the *Fabaceae* family. It was also common for these early helmets to be made of wicker and covered with white cloth to provide coolness to the wearer. While the helmets were more commonly used by HEIC officers rather than crown officers, such eminent Victorian soldiers as Major-General Sir Charles Napier, KCB, were among those who wore these early helmets.

The helmets have come to be known as the "air pipe" helmet because these featured a comb on the top – reminiscent of the comb on cavalry helmets of the era – with a hole at the front to provide ventilation. Few examples of these early helmets are known to have survived.

An early example of the air-pipe helmet which became popular with British officers in India in the 1850s.

These helmets were used during the Indian Mutiny of 1857. British forces used the second pattern 1855 Shako, and forage caps, as the primary types of cover, but during the Mutiny many officers acquired locally made helmets or English felt helmets. These locally made helmets were made of cotton or canvas on a wicker frame. It is likely that these were based on the earliest sun helmets provided for civilian use in the region.

It was following the Mutiny that special headgear and warm weather garments were at last considered and officially approved for use by the British Army. As W.Y. Carman says in "Indian Army Uniforms, Infantry."

> The most striking changes were in head wear, for the full dress shakos and helmets as worn at home were 'out' and a tropical helmet was used for parade and ceremony when the undress cap could not be worn. 'Helmet forage caps' had been sanctioned 18 November 1854, for British officers to wear on parade when white covers were worn by

An example of the white helmet authorized in 1877. This one is circa 1890s.

The interior of the above white colonial helmet showing the plush silk lining. Note the absence of any maker's/outfitter's mark.

the man (sic) on their stiff head-dresses. A painting of an officer of the 22nd Madra Native Infantry, 1856, shows a high crested helmet with a quilted top and the ends of the pugri hanging down the neck. Wicker helmets are noted in the order of 1 April 1859, and a further order, No. 47 of 11 May describes a light felt or wicker helmet which was to be worn with a white quilted cover and a turban or (sic) unbleached muslin. In July it was ordered to be worn on all duties on which the forage cap covered was to be worn. In August 1859, when khaki was worn as a summer dress, a khaki cover was to be worn with the wicker helmet.
A good business appears to have been made in helmets for there were various names like 'Elwood's patent' etc. The permanent choice was the new patent cork helmet ordered for European officers in Native regiments on 5 August 1870. This was to be made by Messrs. Hawkes and Co. of London. It was to be covered with white cloth with a regimental pugri and a gilt curb-chain.

From 1870 a cork helmet with an air vent at the top was issued to all regiments serving in India, and it was this style of helmet that was used during the Abyssinian campaign in 1868, and the Ashanti War of 1874 in West Africa. In June of 1877 a white helmet of similar design was officially authorized for wear by all ranks throughout the Empire.

An example of the Colonial pattern helmet without helmet plate nor any other adornment. Some regiments, such as the 99th of Foot, did wear a helmet plate during the Zulu War of 1879 but this was a very rare practise.

Known as the Foreign Service Helmet, it was made of cork covered in white cloth with six panels. Peaks and sides were bound in white cloth, with a one-inch wide piece of cloth sewn around the headband above the peaks, this being covered by a puggaree, a cloth wrap around the outer headband of the helmet. The puggaree was authorized for certain stations such as Hong Kong, Bermuda and Malta. A zinc button, covered in white cloth, was fitted to the top of the helmet.

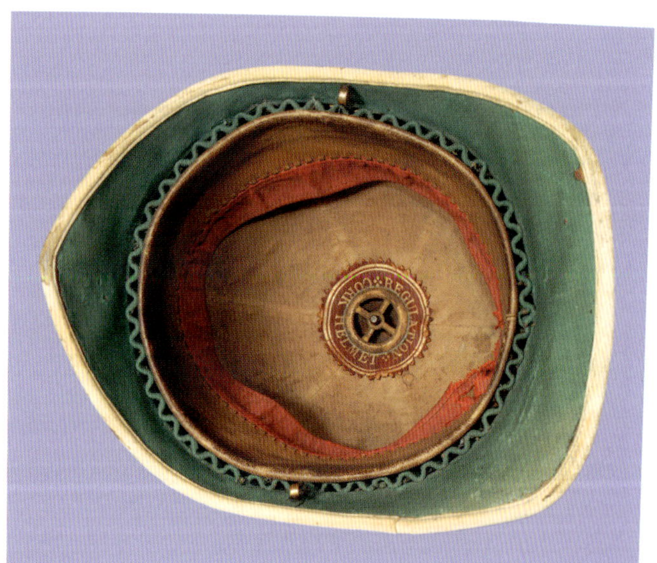

The interior of an officer's 1877 pattern helmet. Note the red silk attached to the leather sweatband and the corrugated ventilation between the sweatband and the shell.

The helmet of Sir Charles Venables-Llewelyn, Glamorgan Yeomanry. This illustrates the gilt chinchain normally worn on occasions requiring full-dress.

Sir Charles Venables-Llewelyn in the uniform of the Glamorgan Yeomanry.

Khaki and Helmet Features

In addition to the helmets, the British forces in India also began to experiment with a new color of uniform. The famous "Red Coat" had been in widespread use since the days of Queen Anne, and while the Victorian Age had been one of strict pomp and circumstance, the British Army was learning many costly lessons of war. Among these was that bright white helmets may look good on parade, but they didn't offer much in the way of camouflage.

The word "khaki" is the Persian word for "dust" and its Hindustani equivalent is "khak", and while serving in India it was common to attempt to break up the white color of helmets, as well as other white garments with the use of mud, tea, coffee, tobacco juice and curry powder. The results varied greatly, and by 1884 a patented dyestuff produced a fast khaki dye, but it wasn't until the second Afghan War in 1885 that a fully khaki uniform replaced the red uniforms and white helmets. The Foreign Service Helmet was mostly produced in white drill until the Boer War but khaki covers were extensively used on active service thereby preserving the helmet for ceremonial duties as well as providing camouflage in the field. For this reason it isn't uncommon to find Foreign Service Helmets from much later than 1885 that are white in appearance.

A Boer War era British sun helmet with the flash of the Royal Artillery, believed to have been owned by a Captain O'Malley.

In 1877 helmets were authorized to have a puggaree for station in Malta, India, Ceylon, Hong Kong, the Straits Settlements, the West Indies and Bermuda, St. Helena, Canada, West Africa and the Cape. The purpose of this cotton cloth wrapping was to help keep the helmet cool. Exactly how well this worked is left to debate, and based on period photos and surviving examples, it seems that considerable license was exercised by troops, especially in South Africa, the Sudan and in Egypt. There are many surviving examples of helmets with and without puggarees, but the 1900 Dress Regulations authorized the use of puggarees for all stations after the Army Order 83 of 1896. These Dress Regulations first described the helmet as the Colonial Helmet.

Depending on where the helmet was used, a helmet curtain may have been made available to provide shade to the neck. This would wrap around the helmet and was tied from the front. As the helmet curtain was not permanently attached few examples have survived and these must be considered extremely rare.

A Colonial pattern khaki helmet of the Boer War period. This example shows a puggaree of 9 pleats with flash and is to the 4th Battalion of the King's Own Royal Lancaster regiment.

Helmet badges and fittings

One major misconception surrounding the Foreign Service Helmet is that these all had the unit plates on the front. No doubt this misconception is due to movies such as Zulu. In fact, most of the Foreign Service Helmets never were fitted with plates or any form of badges while being used by the British Army, and almost certainly not on campaign. One reason for this confusion is that colonial units, including the Natal Mounted Police, Natal Carbineers, Durban Mounted Rifles and Transvaal Rangers did wear plates, and more often spikes. The spike when fitted into an acanthus leaf base, a simple dome or the 4 finial crosspiece base of the Home Service Helmet, screwed into the same threads as the zinc button. Both the plates and spikes were typically issued in white metal or brass, and examples of each are encountered.

The chinstrap used in the field is another feature that has continued to spark debate among collectors. There are numerous photos that do suggest that for parade duties British officers, especially in the years following the introduction of the helmet, opted to wear a spike and chin scales. This would have been in keeping with the style that the officers grew accustomed to, considering that the newly adopted Home Service Helmet featured a brass spike and chin scales. But outside of the parade grounds, the spike certainly wasn't usual for most officers. Additionally while the British Army did not typically wear spikes or plates, the same rules did not seem to

Three soldiers of the Gloucestershire Regiment showing the neck curtain and the famous back-badge awarded for the battle of Alexandria in 1801 (Courtesy Soldiers of Gloucestershire Museum).

Signallers of the 2nd Battalion Gloucestershire Regiment in Mesopotamia 1916-1918. The puggaree on the Wolseley of the soldier on the left is unusual and seems to be a single strip of cloth without pleats (Courtesy Soldiers of Gloucestershire Museum).

This Wolseley is to the 7th Queen's Own Hussars. The flash colours are yellow and dark blue but the yellow has faded to almost white.

A Wolseley helmet to the 8th Hussars featuring a horsehair hackle of the regimental colours.

A 1937 dated Wolseley being to the Royal Engineers complete with slider badge for the puggaree.

apply to the British Marines units, who wore a spike – or a ball for Marine Artillery units – along with the regimental-pattern helmet plate. It was more common for units of the British Army, especially in the later decades of the 19th century, to adorn the unit insignia on the left hand side of the helmet, although there were notable exceptions. This was typically a unit cloth patch, which was often removed from the shoulder strap, and was either tucked into the puggaree or sewn to it.

By the end of the 19th century the Foreign Service Helmet was in use throughout the empire and it was on average liked by the troops. Enlisted men and NCOs were supplied with a helmet, and often times a cover, while officers had to purchase their own helmets. As with the Home Service Helmets, several manufacturers produced the helmets, and thus the collector may encounter subtle – even not so subtle – differences in design. The primary manufacturers and outfitters of these helmets are believed to be the same makers and outfitters of the Home Service Helmet, and included Hawkes & Company of Piccadilly, London; Hamburger Rogers & Sons, London; Humphreys & Crook of Haymarket, London; J.B. Johnstone of Saville Row, London, and Dawson St, Dublin; H. Lehmann of Aldershot; and Samuel Gardner & Co of Clifford St, London.

The Wolseley Pattern Helmet

By 1896 the British Army adopted khaki uniforms as the standard campaign dress and the day of the scarlet uniforms on the battlefield had passed. This was the twilight of the Victorian Age and in 1901, after ruling an empire that stretched around the world the great British Queen and Empress of India died. Throughout nearly every year of her reign British troops engaged an enemy somewhere around the globe, and at the time of her death the 2nd Anglo-Boer War was still being fought. For the British this was a new type of war, and many hard lessons were to be learned. It was also the last war where the British Army used the Colonial Helmet in large numbers.

At the start of the war the helmet was almost universally issued to British forces, while experiments began with other styles of headgear notably slouch hats. But with the harsh realities of trench warfare still more than a decade away, the tropical sun helmet remained a popular style of headgear. This new helmet was the Wolseley Pattern, but despite the name, Sir Garnet Wolseley did not design it. In fact it only took on this name later, in part because it was the preferred style of helmet by this eminent Victorian soldier.

A full-dress Wolseley helmet to the Middlesex Regiment showing puggaree with piping, badge, spike and chinchain.

This helmet had a flatter and wider brim. Period photographs suggest staff and general officers first used it during the Sudan Campaign in 1896, but it became especially popular with other officers, and even the troops toward the end of the Boer War. During the conflict most regimental officers continued to wear the older pattern helmet, no doubt to avoid standing out to the talented Boer snipers.

The helmet was made of cork, covered in white or khaki drill cloth with six seams. The peaks measured 3" wide at the front, 4" at the back and 2" at the side. The helmet was bound around the headband with a khaki puggaree and typically had a brown leather chinstrap. This helmet began replacing the older Colonial Helmet beginning in 1901 and was in general use throughout the empire at the start of World War I, seeing service in the Gallipoli, Middle East and African campaigns. As most of the helmets produced for these campaigns were khaki in color, a cover was never considered.

An extremely rare side hook, under the puggaree, is shown on the Middlesex full dress helmet to accommodate the looping up of the chinchain.

This helmet became a "sealed" pattern in 1899 and referred to in the 1900 Dress Regulations as the Wolseley pattern cork helmet, but initially only for the West India Regiment and the Chinese Regiment and only in khaki, while the '1904 Dress Regulations' state that it "was for service at all stations abroad", later it was also to be of white drill.

As with the Colonial Helmet, the Wolseley was typically worn without spike, plates or other accoutrements unless on full dress occasions. Various band units, and the Royal Marines, continue to use this style to this day.

British regiments were very jealous of their regimental distinctions and one such manifestation was piping to the puggaree. This was in the form of silk pleats or a strip of cloth (woolen or cotton) to the top of the puggaree or to a particular fold of the puggaree. For example the South Wales Borderers regiment had the fourth pleat from the

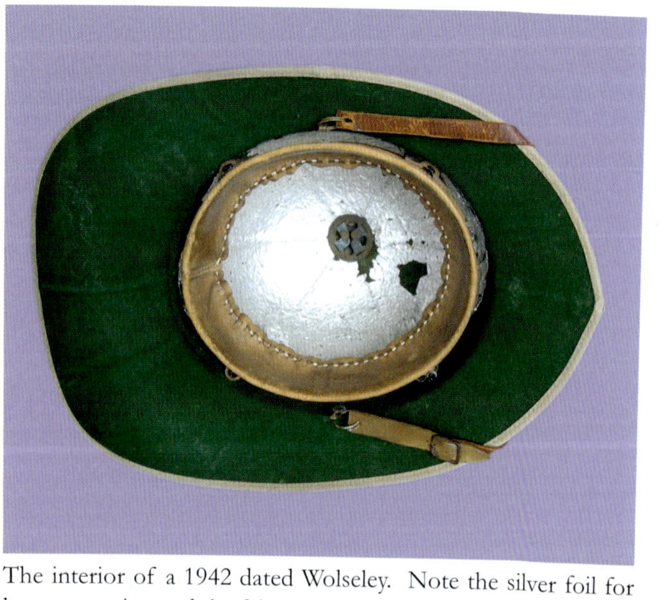

The interior of a 1942 dated Wolseley. Note the silver foil for heat protection and the felt construction. The ventilation button is secured by 4 bent metal tabs because of the felt construction.

A Wolseley to the Royal Berkshire featuring the flash of that famous regiment. This one is attributed to Frederick Rance of the Machine Gun Section and is of WWI vintage.

A Wolseley helmet to the Essex Regiment which was also known as the Purple Pompadours because of the colour of the flash.

A Cawnpore Tent Club helmet made famous by the Prince of Wales' visit to India 1875-76. Used initially by British forces for undress or fatigue duties.

A Bombay Bowler to the Northamptonshire Regiment. The Bombay Bowler does not appear to have any official presence and was a down-sized Cawnpore or Khaki Solar Pith Hat. It was probably a civilian hat commonly used by officers.

top green and the fifth pleat white.

Another form of regimental/battalion or divisional distinction was the flash or patch attached to the puggaree and normally worn on the left side of the helmet. It was not unusual for some units to wear patches on the right hand side or even on both sides. These patches varied greatly in both size, shape and color.

British Sun Helmets of the 20th Century

Much of the empire encompassed tropical regions and that meant a lot of troops needed to wear sun helmets. The Wolseley remained in service throughout World War II, but had been supplemented by the Cawnpore Tent Club hat from the latter part of the 19th century for fatigue

The interior of a Khaki Sola Pith helmet. Note the one inch cross tapes of cotton and also the eyelets, two per side, to provide extra ventilation. It is also interesting to note that period photographs taken during WWII do not show these side ventilation eyelets.

A Khaki Solar Pith Hat to the Royal Armoured Car School. In this example the top "ventilator" button is fixed and has no vent into the shell of the helmet.

A Khaki Solar Pith Hat dated 1937 which officially replaced the Wolseley and Cawnpore for use in India in 1938. This example has the familar Royal Artillery flash of red/blue in a diamond shape. Unlike the Royal Armoured School example above this one has a vent under the top ventilator button.

duties. The visit of the Prince of Wales to India 1875-76 greatly enhanced the popularity of the Cawnpore.

Wolseleys were phased out of use by the Indian Government and replaced by the Khaki Sola Pith Helmet in the early 1930s. Wolseleys continued in use until the late 1940s and were often made of felt or straw because of both the shortage and cost of cork.

In 1938 the Commander-in-Chief India dictated that the Cawnpore Tent Club helmet and the Wolseley were to be officially replaced by the Khaki Solar Pith Hat, both for officers and for Other Ranks.

Wolseleys and Cawnpores currently in use were allowed to be worn but only until December 31, 1938. The Wolseley continued in use outside of India and, in fact, troops departing India for stations abroad were supplied with a Wolseley helmet complete with chinstrap, puggaree and storage bag.

British Commonwealth Countries

It was often said, during the latter years of the Victorian Era, that the sun never set on the British Empire. This was true of course, but due mostly to strategic placement of the Empire's overseas domains, which included India, colonies in Africa, Canada and Australia.

At its height nearly one quarter of the world was under the Union flag, and while this was thanks to the senior service of the Royal Navy, the "Soldiers of the Queen" were responsible for maintaining the peace in far off lands. From nearly its earliest days the British military took an approach that mirrored the Romans, in using various colonial and indigenous forces to help fill the ranks.

For this reason the sun helmet can actually be seen as a quasi-commonwealth contribution, although technically the sola pith helmets of India were developed before the subcontinent was even a Crown Colony, let alone a Commonwealth.

However, for the purposes of clarification, this book will essentially address the various actual commonwealths, and look at sun helmets used by units raised in these far off lands.

Australia

When Australia was settled in 1788 by the British, units of the Regular British Army provided a garrison and performed such duties as guarding convicts, maintaining civil law and order and providing protection from foreign interference. The posting was not popular with the troops because of the inherent boredom, the climate and the sheer distance from their homes and fellow regiments.

The British Empire in 1893 (Courtesy www.youroldbooksandmaps.co.uk).

A white colonial pattern helmet to the NSW Infantry. This example features the "Badge of the Colony" and has the scarlet backing authorised in 1883.

This Wolseley is to Flight Sergeant Thompson RAAF, and was manufacturd in Melbourne, Australia although the maker's name is not legible.

The British Government, always with an eye to costs, soon began to pressure the various Colonial Governments to raise Volunteer forces which would assume responsibility for their own defense. These forces were raised in 1854 in Victoria, New South Wales and South Australia. They were followed by Volunteer forces in Tasmania 1859, Queensland in 1860 and Western Australia in 1861.

These Volunteer forces generally fashioned themselves on British Army lines. Some of these men saw service in the British Army during the Maori Wars and in 1885 a contingent from New South Wales went to the Sudan. In 1870 the British Regulars left Australia and the Volunteers were supported by permanent forces, but in the mid 1880s permanent Militia forces were formed which received payment for service. In 1901, with Federation, control of all Australian forces passed to the Commonwealth of Australia.

An Australian Infantry Wolseley with full ceremonial chinchain and spike. This example is post 1953 as its badge has the Queen Elizabeth II crown.

Used by the Australian Navy since World War II, based on the British Wolseley sun helmet. It features the NCO naval brass cap badge painted blue.

As stated uniforms were closely modelled on their British counterparts with Hussars, Lancers, Horse Artillery and infantry resplendent in scarlet or blue tunics and trousers of blue, brown, gray and rifle green. Headgear included Busbies, plumed helmets, British Home Service helmets, white Colonial pattern helmets with and without puggarees and with helmet plates. It is interesting to note that these white Colonial helmets had both the spike and dome fitting and spike with the Home Service helmet crosspiece depending on the regiment.

The Wolseley helmet was apparently not generally issued to Australian forces, the slouch hat being preferred. However, the Australian Intelligence Corps were authorized to wear the Wolseley Pattern helmet, by order MO 336/1908, with a puggaree of seven folds, the centre fold being pale blue, and with a small regimental badge worn on the front of the helmet.

Examples of the Wolseley being issued to the Royal Australian Air Force and Royal Australian Navy are known to exist and some of these helmets were manufactured in Australia.

A rare Wolseley dated 1913 to the Australian Engineers.

Canada

The North American portion of the British Empire made up nearly one half of the empire's total landmass, yet it required a relatively small local military force to maintain the peace. Throughout the 19th century there were no major uprisings or significant rebellions. However, despite the fact that Canada isn't exactly a "tropical region," the Canadian forces – which included the famous Royal Canadian Mounted Police – did use, at times, a white sun helmet.

The Canadian forces followed the Imperial regulations, but there were some variations. Originally the Canadian military followed the British regulations, which called for the use of the blue/green Home Service helmets. It is worth noting that the Home Service helmet worn by one of the Ontario units in the 1885 Reil Rebellion, marked the only time that the Home Service helmet ever saw action in the Empire!

However, this helmet pattern was only used in Canada

A soldier of The Royal Canadian Regiment on sentry duty in Bermuda, 1915. He wears the short-lived Canadian 1913 pattern Web Equipment. Note the khaki cloth cover to his helmet. (Collection of The Royal Canadian Regiment Museum).

The seven-fold pugaree was approved in 1914 for The Royal Canadian Regiment but evidence exists that shows the The RCR wore it as early as 1913 (Collection of the Canadian Forces Base Gagetown Museum).

until 1886, when it was withdrawn in favor of the white Colonial pattern helmet, which remained in use until it was replaced by the British Wolseley pattern.

The white sun helmet was also a regular part of the uniform of the Canadian Mounted Police beginning in 1876. The "Universal Pattern helmet" was upgraded with the addition of gilt hardware when used as part of the Full Dress uniform, and this was worn without puggaree.

The helmet pattern for the Mounted Police was modified slightly in 1901, but the basic shape remained consistent. It was still produced of cork, covered with white cloth in six seams and bound with buff leather at the bottom. The specifications for the helmet were formalized in the dress regulations authorized by Council on June 22, 1904, and also appeared in the regulations published in 1909.

Wolseley helmet as worn by the Royal New Brunswick Regiment. This, and the helmet worn by The RCR, have plastic spikes and locally-produced chin chains. The Wolseley helmet is the full dress headdress for all regiments and Corps of the Canadian Army who do not have an approved headdress i.e. Busby, Bearskin, etc. (Collection of the Canadian Forces Base Gagetown Museum).

Additionally, the most significant difference with the Canadian white Colonial pattern helmet is that unlike those used by the British regiments in the latter years of the 19th century, the Canadians wore these while serving at home, complete with metal fittings including badges, and even colored puggarees.

Made by Hawley products Canada, Brantford, Ontario. It displays the C/Arrow Cdn Army property mark. The issue of the Hawley "Tropper" sun helmet was universal in Canada. Although known for its snow and ice the country is extremely hot during the summer months.

By the first decade of the 20th century the transition from the colonial pattern to the Wolseley was complete, and again the Canadian helmets often had more flair than those used by British regiments. This was because in many cases the helmets were used as much as dress helmets as for use in the field. This helmet pattern remained in use during the interwar period, although the Wolseley was gradually replaced by the Bombay Bowler pattern.

The Wolseley was also adopted by the Canadian Mounted Police between 1911 and 1923, and was first described in the Dress Regulations of the Force updated in 1923.

With the outbreak of World War II the Canadian Army also looked south to the United States, and produced its own version of the pressed fiber sun helmets. These were made by the Canadian division of the Hawley Products Company, and feature a slight variation from the American pattern. The most distinguishing feature is that these helmets feature three rows of ventilation holes with eyelets on each side, and typically lack the front vent hole, which allowed for the placement of a cap badge.

British Caribbean

There have been many regiments raised in the British

Based on the same basic pattern as the American pressed fiber sun helmet; the Canadian versions, which were produced by Hawley Product Company Canada, had minor differences. This example is unique in that it has a real top ventilator and three holes in the back of brim, possibly for attaching a neck screen.

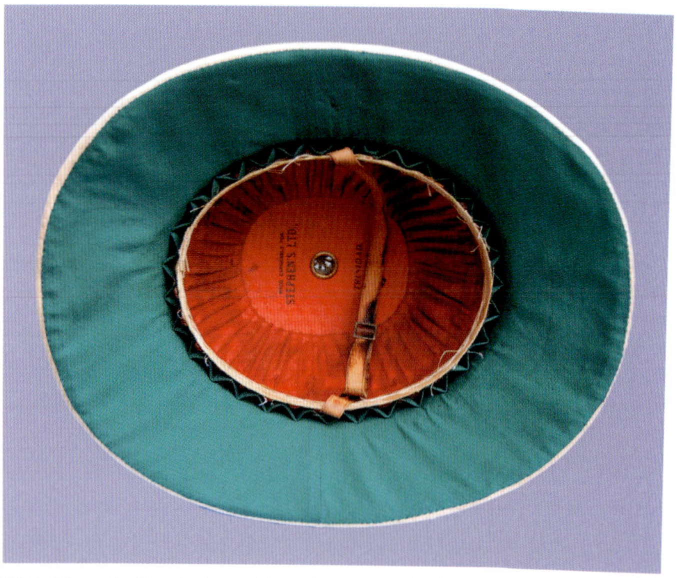

This West Indies helmet liner is unique in that instead of leather it is made of woven straw.

Caribbean and possibly the first was the Barbadoes Regiment of Foot (Bridge's) raised in 1667. Perhaps the most well known is the British West India Regiment which was last reformed in 1915 from the various British colonies of the Caribbean. It was a hostilities only regiment and was disbanded in 1921. This regiment was unusual in that it formed part of the British Regular Army.

The various commonwealth military forces of the British Caribbean including those of Jamaica, the Bahamas, Barbados and British Guiana used colonial pattern helmets throughout the latter years of the 19th century, and adopted the Wolseley pattern helmets in the years leading up to World War I.

Even today these nation's police forces continue to use Wolseley style helmets – often locally produced. These helmets often lack the puggaree but are otherwise similar to the basic Wolseley helmet pattern used throughout the empire for more than 80 years.

A pre-World War II tobacco card from Player's Cigarettes shows a British Guiana Militia Company Sergeant-Major, wearing a white Wolseley pattern sun helmet, with brass spike, front badge and chinchain.

A World War I era sun helmet based on the Indian patterns. Locally made in the West Indies it features the puggaree of the West Indies Regiment.

Chapter 2

French Helmets - La Casque Coloniale

For approximately 100 years, from the second half of the 19th century to the 1950s the French Empire ranked second only to Britain in size. Unlike the British system of governing distant colonies the French relied on a more centralized system. Despite this fact a variety of French units served overseas, and these included Métropolitaine troops, private trading company soldiers and local militias. However, following the downfall of the French Second Empire under Emperor Napoleon III in 1870, the role of empire building at once fell to the French navy – which had been a vigorous proponent of an overseas empire since its formal founding by Colbert in the 17th century.

During the era of the Third Republic there was actually an objection to a colonial army for the tropics – due in part to the law that French conscripts could not be forced to serve outside Europe. At the same time there was a feeling that volunteer armies equated an elite guard – which had aided two Bonaparte emperors to the throne. For this reason for much of the final half of the 19th century the empire building was done by Troupes de Marine, so it was marine units that marched across the vast Sahara Desert in an effort to build the French Empire. Towards the end of the century and in the years leading up to World War I the role of garrisoning the empire was shared with the Métropolitaine. Eventually this French Marine Infantry fighting force was renamed Troupes Coloniales in 1900. However, to add to the confusion for historians and collectors alike this should not be confused with the Armée d'Afrique, which was actually made up of North African volunteers to the French Army.

This is worth noting because the early colonial pattern French sun helmets of the 19th century typically bore a brass anchor on the front. These helmets were used by the Troupes de Marine – the French Marine Infantry – and should thus not be confused with "naval" helmets.

French Marines with the officer wearing the anchor badge whilst the men have no such adornment.

An example of the brass anchor worn by the French Marine Infantry.

The "Sugar Loaf" Helmets

From its introduction in the 1870s through the late 1930s, there were two basic models of French sun helmets. These included the two colonial pattern helmets, known as the Model 1878 and the Model 1886 – which derived from the early pattern. These helmets, which were both likely inspired by the British Foreign Service helmet, and were also both known as "sugar loaf" helmets due to their tall slender appearance.

The Model 1878 was made up of six panels of cork in the dome of the helmet, along with an additional six panels around the skirt curtain of the helmet. The helmets were covered with a heavy white fabric, which was typically "painted" with blanco to re-whiten the helmet as it showed dirt. At the top of the helmet was worn a ventilator cap. The French Troupes de Marine typically did not wear spikes or other accoutrement such as chinchains. However, on the front of the helmet was worn a brass anchor – the symbol of the French Marine Infantry. Again, these should not be confused with "naval" helmets.

The Model 1878 helmet remained in service for many years, and thus in period photos these helmets are often seen along side the later Model 1886 helmet. The evolution from the Model 1878 to Model 1886 is unique in that the latter helmet features far less flair in either the front or rear brim of the helmet. As with its predecessor helmet it also featured a six-panel construction with ventilator cap on the top, and a French Marine Infantry badge on the front.

Neither model was ever issued with a puggaree, and it seems that the French military never even considered utilizing such an addition to the helmet. However it was not uncommon for officers to occasionally add a hatband around the helmet.

Both model helmets featured a simple liner; usually leather but other fabrics and materials have been encountered – including those that may have been "field replacements." To provide ventilation the liners were mounted to leather spacers to provide airflow around the wearer's head. Some surviving examples even feature red silk, suggesting that these may have been private purchase officer's helmets.

Beginning in the 1880s the French began to issue covers for the helmets, and these were originally light blue to correspond with the Troupes de Marine tunics. The covers for the helmet appear to have been consistent throughout the latter part of the 19th century, and it is not believed that there was a specific pattern for either colonial helmet.

As with the British tradition to dye helmets in the field, some French colonial pattern helmets clearly show signs that this happened. It is not believed that any other fabric color was used to produce the helmets however.

Both patterns of helmets were fabricated in Paris as

The Model 1878 features a clearly pronounced visor and the nape sheild. This pith helmet features the insignia of the French Marine Infantry, and were used throughout the French Empire in Africa and Indochina.

This model 1886 pith helmet features the insignia of the French Marine Infantry. These helmets were used in various French colonies, including Africa, and were used by the 9th and 11th Marine Infantry that fought in China during the Boxer Rebellion.

well as Marseille. The latter was also the typical departure city for troops headed to both Africa and Asia. Multiple firms in Marseille produced the helmets, and many examples feature stamps indicating the maker of the helmets.

In the years leading up to the First World War the role of guardians of the empire fell to more than just the French Marine Infantry. Thus later examples of the Model 1886 pattern – likely made after 1900 – do feature the bronze grenade of the Gendarmerie. This of course was the same basic symbol that would be later seen on the

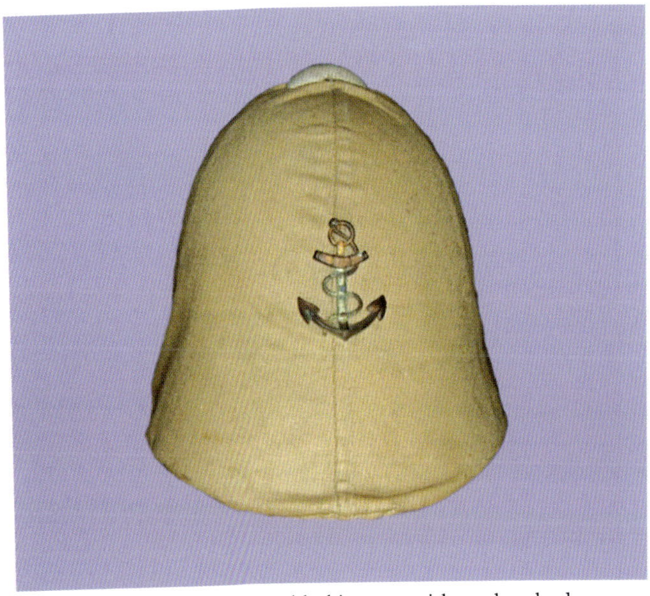

An 1886 helmet featuring a khaki cover with anchor badge.

An 1895 poster, from the cover of the book La Guerre A Madagascar, depicting the French invasion of Madagascar.

The liner of an M78 helmet.

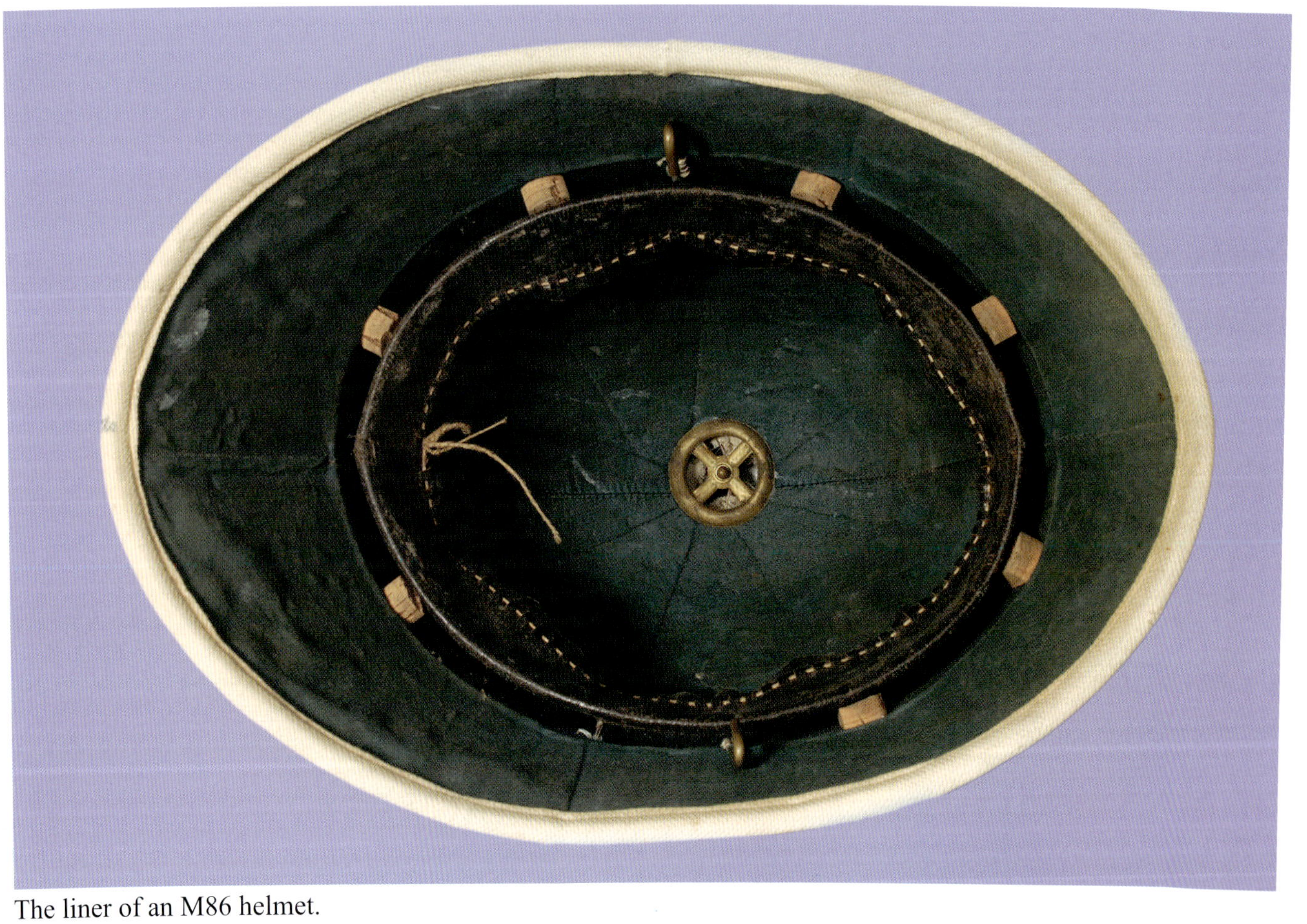

The liner of an M86 helmet.

French Model 1915 "Adrian" steel helmets.

Other steel helmet badges may be encountered on these helmets, but whether these were added to "spice up" a helmet was left to one's own discretion.

The Model 1931 Sun Helmet

The colonial pattern sun helmets of France remained in service longer than the British Colonial pattern that served as the inspiration, and the so-called sugar loaf helmets were only replaced by a more modern helmet design more than a decade after the First World War.

This was the Model 1931, which was produced in a number of variations, and saw service throughout World War II and in the post-war colonial uprisings in the 1950s.

French Model 1931 Tropical Helmet of the Regional Gendarmerie. A variation of the Model 1931, this example features the dark blue headband indicating an auxilary Gendarmerie. These troops saw service in Algeria during World War II.

The helmet has the characteristic look of the so-called "safari" helmets of the 19th century. The design of the original Model 1931 helmet, based on the British Wolseley, is quite simple, yet numerous variations of this pattern exist, which is not surprising given that this helmet was used for nearly 30 years.

The earliest variation of the Model 1931 featured four cloth panels over a rolled cork dome. As with the previous colonial patterns the helmet was capped with a simple ventilator. Most early examples also feature a single piece of cloth covering the brim of the helmet. Numerous colors of cloth were used in the 1930s and thus helmets could range from pure white to khaki.

Among the variations were helmets that had a less even brim, where the rear nape flared out almost resembling the British Wolseley pattern helmet. These were believed to have been produced in the years immediately leading up to World War II.

A more common pre-war variation saw a more sloped brim, along with a six-panel construction instead of the previous four. Additionally the French makers added side ventilation holes – and the pre-war number is believed to range with either one on each side or two, while the post-WWII examples may feature as many as three side ventilation holes. All of these include metallic grommets, either of painted steel or brass.

A Gendarmerie grenade badge worn on the model 1931 helmet.

As with the previous colonial patterns the various Model 1931 helmets were fabricated in Paris as well as Marseille, including those that were used by the French Army in Indochina in the early 1950s.

The model 1931 tropical helmet was used throughout the French colonies, including those in Africa, the Middle East, South America, the Pacific and French Indo-China. This example features the standard marine anchor.

During World War II a variety of badges and insignia were used on the helmets, and these range from Vichy French to Free French. Likewise during the war the Free French military relied on a variety of equipment from various other powers – notably the British and Americans. As such, the French forces in Africa and the Middle East did use surplus British Wolseley helmets, South African helmets, Dutch and even American sun helmets. Thus it is possible to encounter these with French insignia, but as this wasn't standard, collectors should be wary of potentially "made up" examples.

This style of helmet was based on the French Model 1931 tropical helmet, but is made of a lightweight pressed fiber with a waterproof canvas cover. These were locally produced in French Indochina, and used as private purchase officer's helmets. These feature a plastic liner system, which would be later used with the North Vietnamese Army sun helmets.This example was made in Saigon, and has a maker's tag from Nón Rang Dong, BVK. The original owner's initials(?) is written in pen: "VAN 3-6-54."

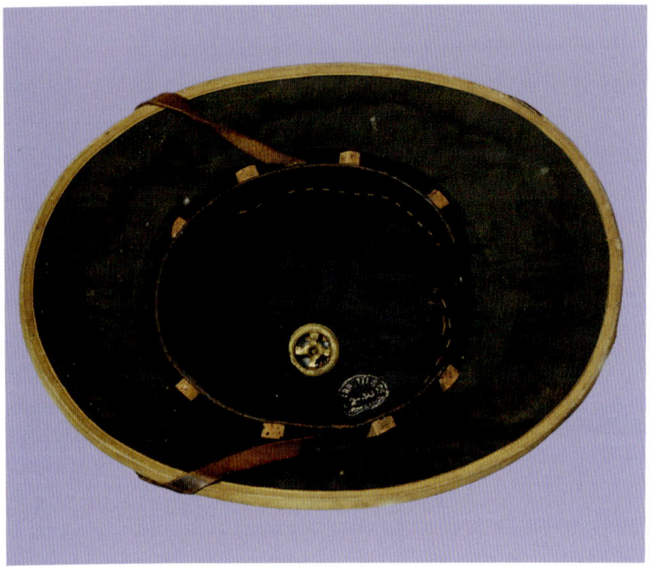

The interior of an M31 showing the cork spacers around the headband which provide ventilation.

Additionally with the fall of France in 1940, Germany captured large stockpiles of French equipment including helmets, and it is believed that some Model 1931 helmets may have been used by Kriegsmarine (naval) and Police forces in the Mediterranean

Post World War II Helmets and Export Use

Following the Second World War, France desperately attempted to retain its dwindling colonial empire, including French Indochina. While steel helmets were widely used in the campaigns in Asia, as well as in North Africa, officers and other military officials continued to look for the traditional sun helmets.

In the early 1950s there were even locally produced Model 1931 helmets made in French Indochina, including in Saigon. It is believed that these may have been the forerunner in the use of the rubberized liners that were later used in the North Vietnamese sun helmets.

Additionally, it is worth noting that the French Model 1878 and the Model 1886 were widely exported. Officers from the armies of Belgium, Portugal and Spain relied on these patterns of helmets –notably the later M86 version – as private purchase items. Even today a version of the Model 1878 helmet is used by the Compagnie des Carabiniers du Prince (Prince's Company of Carabiniers), the ceremonial unit of the military force of Monaco. This is believed to be among the very last of the colonial pattern helmets of Europe still used by a military force today.

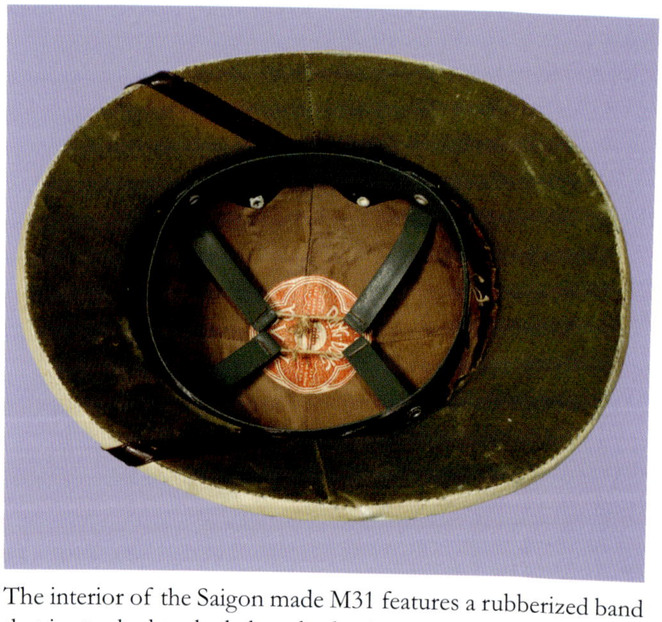

The interior of the Saigon made M31 features a rubberized band that is attached to the helmet body via two sets of rivets on each side. These rivets double as the ventilation hole grommets for the helmet, which must be necessary as the top ventilator comb is for show only. This system of liner was likely developed in Indochina and was the type used in the sun helmets of the People's Army of Vietnam (the North Vietnamese Army).

Summer dress helmet of the Compagnie des Carabiniers du Prince (Prince's Company of Carabiniers), the main ceremonial unit of the military force of Monaco. This pattern helmet is based on the French Model 1878 sun helmet, and remains in use today as the white summer helmet.

The white summer uniform of the Compagnie des Carabiniers du Prince is topped off with a white sun helmet.

Chapter 3

German Sun Helmets - Imperial Tropenhelme to Afrika Korps

Germany was a late arrival on the colonial scene, only becoming a unified nation in 1871. Prior to that time the various German states were too small, too landlocked and simply too involved in affairs in Europe to even consider overseas colonies. Even with the unification of Germany the opinion was that colonies were unnecessary. However, in 1884 Chancellor Otto Von Bismarck, who had been opposed to colonial expansion, suddenly changed his mind.

An undated photo shows a German soldier wearing the early colonial pattern helmet that was used throughout the Imperial German Empire until the end of the First World War.

Germany established colonies in regions that had already been part of German trading interests, and suddenly the nation had an overseas empire! These included Cameroon and Togo in Central West Africa, German Southwest Africa at the Cape, and most notably German East Africa on the Indian Ocean. The last was the scene of the most German resistance in Africa during World War I as Lt. Colonel (later General) Paul von Lettow-Vorbeck lead a campaign up and down East Africa keeping thousands of British, Indian, South African, Belgian and Portuguese forces tied up pursuing him!

Additionally in the years leading up to World War I, the Germans established colonies on islands throughout the Pacific and even a small trading colony at Tsingtao in China. What is both fascinating, and somewhat confusing, is that these colonies were independently managed to the point that many had their own local armies with equipment completely specialized for that particular small German corner of the world.

Numerous efforts have been made to chronicle the various colonial German equipment and headgear, and for the purposes of this book only an overview on the subject is possible.

Imperial Navy Sun Helmets

Officers of the Imperial navy were first issued sun helmets in July of 1885, tropical helmets were authorized for officers of all categories, officials of officer rank, deck officers and naval cadets. This was described as a tropical helmet made from Indian reeds or cork with a white cloth linen cover. These helmets had a wide gold edging around the hatband, while narrower for Deck Officers. Photos in Schiers' book show that at least some Naval helmets had the imperial black/white/red cockade on the front.

In 1891 a khaki helmet for other ranks, also made from Indian reeds, was introduced. Additionally, various alterations were made to the pattern, and these were mainly to the ventilation in the 1890s. Moreover,

A Navy (Marine) engineering officer's helmet which has a double silver bullion cord around the base of the crown with a Reich Kokarde (Courtesy www.derritmeister.com).

Note the broad rear brim of the Navy (Marine) engineering officer's helmet (Courtesy www.derritmeister.com).

production transferred from the A. Emter firm in Berlin to the H. Pfeiffer company in Hamburg in 1894/95. Still made from reeds, although cork was also likely used, these sun helmets were used by the navy in the Indian Ocean, in New Guinea, Africa and China.

The navy received the Bortfeldt helmet in 1902. This remains one of the most interesting sun helmets used by Imperial Germany. It was a tall colonial pattern helmet that was covered in a heavy khaki cloth. What was unique is that this helmet had a long rear brim that actually folded up, supposedly for ease of firing while lying prone.

A large neck curtain was quite popular with these helmets and provided an adequate amount of shade from the sun. Surviving examples with the neck curtain are extremely rare.

An example of the Bortfeldt helmet which featured the folding rear brim, supposedly for ease of firing while in a prone position. Ludwig Bortfeldt of Bremen designed this helmet, and his firm was the only manufacturer of the helmet.

The Bortfeldt helmet with the folding rear brim clearly visible.

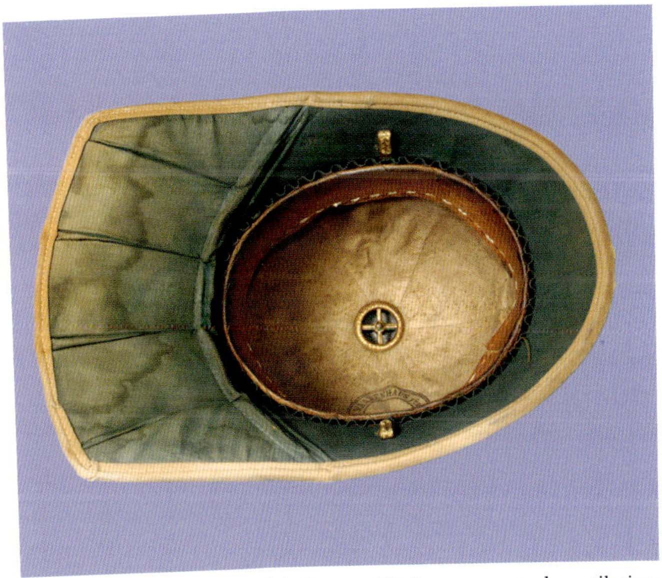

The interior of a Bortfeldt helmet with the corrugated ventilation strip between the shell and the headband clearly visible, as are the chinstrap hooks.

Schutztruppe Helmets

While Germany's colonial history in Africa was full of bloodshed, the actual acquisition of these colonies had been fairly peaceful, and the territory it acquired was determined at the conferences in Berlin between the great powers, and to a lesser extent through treaties with the local powers. German East Africa was acquired through a treaty signed with the Sultan of Zanzibar, which caused some tension with the British. This led to a treaty between the two powers in 1885 that granted Germany the entire hinterland of Zanzibar as far as Lake Tanganyika. In these lands would take place the wild chase of von Lettow-Vorbeck on land, and the almost surreal naval battle on the Lake itself.

Unlike Britain and France, Germany never stationed regular army units in its colonies. Nor did it build up large colonial armies for use in other distant conflicts. Instead it administered its colonies on what could only be described as a shoestring budget. But in fairness the colonies actually never became the source of income that the government in Berlin had hoped for. In fact, trade with Norway accounted for far more revenue than any of the German colonies at the turn of the century.

The larger German colonies were protected by the regular colonial troops known as the "Schutztruppe." But despite the almost ominous sounding name, there were actually very few of these troops stationed abroad at any one time. These men, while few in number helped maintain "order" over a vast amount of land.

German East Africa

From 1889 onwards in German East Africa officers and NCOs of the Wissmanntruppe, a police force made up of Sudanese and Somalians, wore white colonial pattern sun helmets. While some had khaki covers, others were adorned with Imperial cockades at the front, and some were even worn with a puggaree. The Schutztruppe (Protection Force) was formed in 1891 from the Wissmanntruppe, and the helmets retained the same white

An 1897 illustration of German Schutztruppe in German East Africa. Note both the white sun helmets with plates and the khaki sun helmets with cockades on the front.

tall design but with imperial cockade, brass eagle with spread wings, and detachable spike with cruciform. A gold cord was worn around the base by officers. Furthermore, in the years leading up to World War I, khaki helmets were also worn and khaki covers were worn on white helmets.

In 1913 new regulations called for the white helmet to be replaced with a new lower design, "safari" style helmet, which when worn on parade utilized a white cover. At this time a three color cord for NCOs was also introduced.

Many variations of the basic shape of these helmets have been seen, suggesting that multiple manufacturers supplied helmets for the Schutztruppe. During World War I a number of British Wolseley helmets were also known to have been used.

German South West Africa

Tropical helmets were introduced to the forces in German South West Africa in 1892. These were also of the typical colonial pattern, and were usually covered with khaki cloth, with white metal imperial eagles. Photographic evidence suggests that these Wappen plates, the state or colonial coat of arms, had wings spread for officers while folded for other ranks. These helmets were adorned with cockades and white metal spike with cruciform base. This pattern helmet remained in use until it was taken out of service in the mid 1890s. The Schutztruppe of German South West Africa did not wear sun helmets thereafter, and instead favored the German slouch hat.

Cameroon

Beginning in 1896 the German forces in Cameroon were authorized to wear the same helmet as was worn in East Africa. This included the new 1896 Schutztruppe regulations, which included the new helmets of the lower design, in white or khaki, and without eagle or spike. These retained the cockades, while officers wore a silver cord around the hatband, and NCOs wore a twisted cord in Imperial colors. As with the other Schutztruppe helmets many variations of the shape of the helmet have been seen.

Throughout the German colonies there seems to have been considerable overlap in the use of helmets, and this is especially noteworthy during the First World War. During this time the Germans took to dyeing the white uniforms to a light brown shade with tea or local plants, and this practice may have been carried over to the white helmets as well. But it has been suggested that the practice was limited with helmets as the cork tended to disintegrate in boiling water!

Polizeitruppe Sun Helmets

The Polizeitruppe German officers and NCOs in East Africa, Cameroon, Togo, New Guinea and Samoa wore white tropical helmets with a small brass imperial eagle above the imperial cockade. From 1912 onwards German Polizeitruppe NCOs ("Wachtmeister") were also authorized to wear a black/white/red cord around the hatband. There were many variations in style and many manufacturers.

Furthermore, the Landespolizei in German South West Africa were issued sun helmets primarily in two areas, the Caprivi Strip and the diamond mining area South of Lüderitzbucht. Their helmets were also white, with the Imperial cockade and brass Imperial crown above it.

Army Officers Seconded to the Colonies

Regular Imperial Army officers were occasionally seconded to the colonies, and these troops wore the

An obviously staged shot reveals the neck curtain that was used with the various colonial pattern helmets.

Schutztruppe uniform but with army insignia. These troops typically wore the basic patterns in use at the time, with a few exceptions. Their sun helmets were adorned with both the Imperial and their own army unit's state cockade below it.

This helmet, which features both the state and Imperial cockades was possibly used by an officer seconded to the colonies, although the naval anchors either were a later addition or suggest service with one of many naval units that fought in East Africa.

Colonial Officials

The doctors, paymasters, gunsmiths and other support personnel that accompanied the Schutztruppe wore the same basic white or khaki Schutztruppe helmets with the Imperial cockade. Above the cockade some officials wore insignia in the form of a brass or white metal imperial eagle or shield with different insignia depending on their speciality. Officials of officer rank wore gold or silver cords around their hatbands – the color again depending on their speciality trade, as with the Schutztruppe from 1913 officials of NCO rank had the three color hatband. And while conforming to the basic regulations, some officials seem to have also worn a hatband – usually in the same colors as the hatband on their field cap – on their sun helmet.

An undated pre-wartime photo shows a number of German troops wearing neck curtains with their colonial pattern helmets. This photo was likely taken in German East Africa.

Governors and Diplomats

German diplomats and colonial governors were also entitled to wear generally elaborate tropical uniforms, and as a result their tropical helmets were white with brass fluted spikes with cruciform bases and Imperial eagle Wappen. Early helmets even had brass chinscales, along with a large Imperial cockade on the chinstrap boss on the right side. After 1891 these were modified to a small Imperial cockade under the Wappen along with a gold cord around the hatband.

For the Kaiser's visit to Palestine in 1898 a special tropical uniform was designed for him along with his retinue. These tropical helmets were khaki, to match the uniforms, and were adorned with either the Prussian guards eagle, or a Prussian guard star Wappen (for the Leibgendarmerie bodyguard). These were further adorned with large Imperial and Prussian cockades on the right and left sides respectively, as well as a brass spike on cruciform base and were worn with neck shade.

Seebatallion Helmets

In addition to the Schutztruppe the German overseas military forces consisted of three Imperial German Marine Infantry Battalions (Seebatallion). These were formed from the Prussian Seebatallion, which was raised in 1852 to protect the coast and guard naval bases. From the 1880s onwards the Seebatallion units were deployed abroad much like the U.S. Marines or British Royal Marines.

In 1897 Germany acquired the Port of Tsingtao, which was then home to the German Navy's East Asia Squadron, and it was garrisoned by the III. Seebatallion.

These helmets were typically of the same shape and patterns as those used by the Schutztruppe but featured a Wappen frontplate with a gilt eagle and anchor device. Usually below the special plate was worn the Imperial cockade. III. Seebatallion, which was formed in Tsingtao, China, was first authorized to wear a white sun helmet, which was made of Indian reeds or cork.

The badge of the East Asiatic Infantry.

Two privates of the East Asian brigade.

Originally produced in white, tropical sun helmets for the Seebatallion were authorized as khaki for other ranks in 1905, but khaki had already been used in China in 1900 and German South West Africa in 1904. Period photos from the Boxer Rebellion in 1900 do show khaki helmets being used. Additionally, the eagle, which had originally been authorized for wear in white metal was changed to bronze in 1900.

From 1904 onwards the Seebatallion also began to wear the Bortfeldt pattern helmet with the folding neck.

East Asiatic Helmets

Originally sent to China in the summer of 1900 to combat the Boxer Rebellion the East Asian Expeditionary Corps initially wore straw hats with their summer uniforms, but in late 1900 this was changed to a tropical sun helmet of the colonial pattern.

It was a khaki covered cork helmet, adorned with an Imperial eagle – usually in brass, although some arms of service had white metal – along with a hatband in arm of service colors. These included white for infantry, red for cavalry and staff, black edged in red for artillery and engineers. These were usually issued with a removable neckshade.

Bortfeldt produced these helmets, and it appears that some variations may have existed including those worn with a spike instead of a ventilation top.

The East Asian Expeditionary Corps was cut to less than half size in 1901 when the immediate danger of the Boxer Rebellion was suppressed. However, the new

An East Asian colonial pattern helmet with a very rare neck curtain which extends well down the wearer's back (Courtesy www.derritmeister.com).

smaller unit, the East Asian Occupation Brigade continued to wear their khaki tropical helmets until 1904 when a new field grey Bortfeldt helmet was introduced to be worn with the new field grey summer uniform. This helmet was similar in style to the helmet it replaced, complete with Imperial eagle, hatband and cockade but slightly smaller in overall size. In 1909 the last East Asian army troops were withdrawn to Germany, and thus many of these helmets returned home, which is why these helmets typically are found today in Germany rather than in the Far East.

One unique variation used in the Far East was the Tropentschako, a tropical shako that was made in the same basic shape as the Jäger shako. However, it was made of cork and with the same decorations as the tropical helmet. The East Asian Jäger troops wore this helmet with a green hatband before they were disbanded in 1901. Additionally the East Asian Train troops also used this Tropentschako with a blue hatband.

Another variation was the uniform of the commander of the East Asian Expeditionary Corps (and the overall allied commander in China) Field Marshall von Waldersee. He initially wore a helmet of the same design as that worn by the Kaiser's entourage in Palestine in 1898 as described above, later photos show him in a standard Bortfeldt helmet with a white hatband

Sun Helmets in Palestine and Macedonia

Stocks of the old East Asian 1902 khaki Bortfeldt helmets were later used during the First World War by the Asienkorps in Palestine and also by German troops in Macedonia. These often retained the imperial eagle, arm of service hatband, and pinned cockade, as well as the neckshade. It also appears that the cockades, eagles and hatbands were as often as not removed. The Asienkorps stopped wearing these helmets in the field soon after their deployment as these helmets looked too similar to the British sun helmets, and caused friendly fire incidents.

Sun Helmets of Austria-Hungary

Following Germany's unification, and the ratification of the Dual Alliance in 1879 Austria-Hungary had been its closest ally. The two empires were linked in a partnership built around the same native language and a similar culture. For this reason the brief history of the Austrian sun helmets is contained in the chapter on German helmets.

The unique tropical shako used in the Far East by German forces.

A woven straw helmet of the type used by the Austro-Hungarian forces in the Palestine Front. These helmets were reportedly made by the Viennese firm P. & C. Habig.

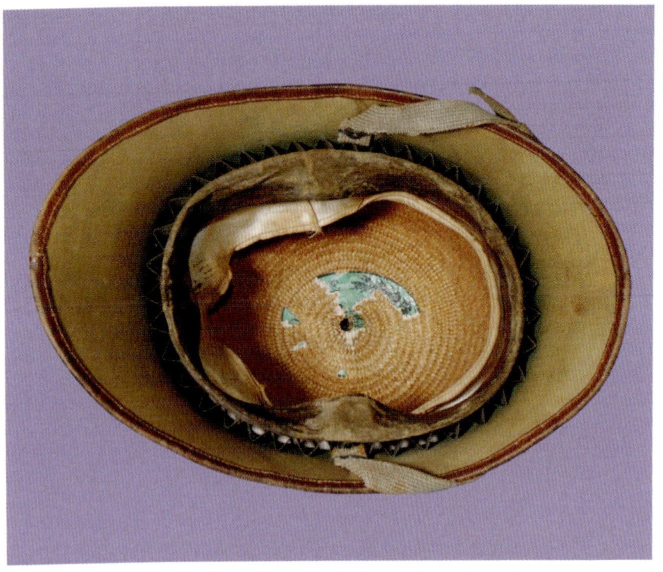

The interior of the Austrian sun helmet featuring the corrugated ventilation between shell and headband and with the chinstrap permanently fixed to the helmet shell.

Despite the fact that Austria lacked an overseas empire, the nation often sent its troops to distant (and often tropical shores). This included patrols in the Pacific Ocean, an 1897 peacekeeping mission to Crete, a small role in the Boxer Rebellion and most notably the "Orientkorps" was sent to aid in the defense of Gallipoli and later the Palestine Front during World War I.

During the Crete Peacekeeping mission the Viennese firm P. & C. Habig produced a number of white sun helmets, and these were likely made of cork. When World War I broke out the sun helmets were all but forgotten within Austria, as these were seldom used except by the Foreign Service and to some extent by the navy. Once again the firm of P. & C. Habig was called to produce the helmets. These were distributed to the troops heading to aid the Ottoman Empire in Palestine. During the First World War a number of patterns were produced, with most retaining the basic colonial pattern shape. While cork was initially used, and covered with a variety of white cotton or linen cloth, as the war went on helmets were made of woven straw that was then covered in khaki-colored cloth. It is worth noting that the basic chinstrap has been called in some sources as being of the same pattern as used for the "Austrian steel helmet" but based on the fact that the earliest pattern sun helmets were made as early as late 1914 suggests that it was the sun helmet chinstrap that was used for the steel helmet.

While many helmets were issued without any insignia or device, later helmets featured the Austrian cap rosette. Those from before late 1916 feature the rosette with the initials "FJ1" (Franz Joseph 1), while later it was a simple "K" (Karl).

Additionally it is likely that captured helmets from Gallipoli were also used by the Austrians in that campaign, and later on the Palestine Front.

German Sun Helmets Between the Wars

One common misconception about the sun helmets used by Germany is actually that sun helmets were not used during the interwar era. In fact this is not true.

While Germany had lost its overseas empire as part of the terms of the treaty of Versailles, the Weimar Republic did maintain a navy. In early 1925 the light cruiser Emden was completed; the first new warship built by in Germany after World War I. It was used primarily as a training vessel, and made several cruises into the Atlantic, the Pacific and throughout the Mediterranean. It was commanded briefly by future admiral and final leader of Nazi Germany Karl Dönitz in 1935.

Naval tropical helmets showing the SMS Emden tally and the faux puggaree.

Along with the K Class cruisers Königsberg, Köln and Karlsruhe, the Emden was one of a handful of warships built prior to the rise to power of the Nazis in 1933. During this time the German navy obtained small quantities of sun helmets for overseas voyages. These helmets were likely produced in Germany, and were made of cork and covered in canvas that was whitened with lime paint. These featured a faux puggaree around the

Sailors of the SMS Emden wearing the tropical helmet during one of its cruises to the Pacific in the early 1930s.

An Afrika Korps first pattern (second version) canvas covered sun helmet showing the army shield with Nazi eagle insignia on the left hand side.

The reverse side of the Afrika Korps canvas covered sun helmet showing the national tri-color shield on the right hand side.

outside of the helmet, and over this was usually worn a black band with the name of the vessel. Inside the helmet had a red and green lining.

Photographic evidence suggests that these helmets were worn with and without a national tri-color cockade, while surviving examples are seen with a gold Nazi-era Naval eagle. This suggests that the helmets remained in use after the Nazis came to power. There are no figures on how many helmets were made, or what firms produced them.

Helmets of the Third Reich

German World War II sun helmets are typically referred to as "Afrika Korp" helmets, but in fact the sun helmets – which offered practically no ballistic protection – were used throughout the Mediterranean theater of operations during the Second World War.

Numerous variants of sun helmets were tested between 1925 and 1940, but as the Reichsmarine, the naval brand of the Reichswehr under the Weimar Republic were originally the only ones using or needing the helmets, development was slow. One pattern helmet was developed in 1937, but as it was made of a molded material, it proved to be unsuitable and wasn't produced in great numbers. Those examples that were already produced were eventually given to U-boat captains, and thus few survived the Second World War.

The typical German sun helmet of the Second World War was interestingly enough developed just in time for the campaigns in Africa. This has always been a curious turn of events, but one argument is that Germany's territorial demands following victory against the Allied Powers likely would have been the restoration of its former African colonies, and possibly additional concessions in tropical regions. Whatever the reason one fact is clear, a lot of sun helmets were produced, despite the other fact that it seems that the troops never really liked these helmets all that much.

The basic shape of the German sun helmets was of the "safari" pattern and the helmet had a silhouette that was reminiscent of the British Wolseley or Italian Model 1928. Two main patterns of helmets were produced, and this has added some confusion. While it does seem that the first pattern was produced in two main varieties somewhat consecutively, with the B version eventually replacing the A version, there was likely overlap during the production life. Additionally, the second pattern was actually produced independently and does not seem to be a later version, but rather just a different pattern helmet

The Canvas Covered Sun Helmet

The so-called first pattern helmet was made of pressed cork or pressboard, and covered with fabric in four sections. A variety of colors of fabric were used, including white, khaki, light green and even bluish gray. The brim of the helmet circled the dome, and was covered with the

A first pattern canvas covered sun helmet, which has the distinguishing 6-panel construction. This example features the Luftwaffe insignia of the eagle in flight.

same colored cloth cut into a single piece. These helmets had no puggaree, and instead a single strip of cloth circled the base of the dome.

The inside of the helmet dome was lined with red or reddish orange cloth, and the there were cork spacers to provide ventilation around the leather headband. The chinstrap was affixed to the helmet via a rivet on the right side of the helmet while a hook for the other end of the strap was affixed on the left hand side of the helmet.

A common variation of the first pattern, sometimes confusingly called the second pattern helmet, is one that consisted of a six-panel construction instead of the previous version's four-panel construction. The brim was also covered in two panels of cloth, and leather spacers were used instead of cork to provide ventilation around the headband. There were also variations that featured a zigzag pattern of pressed material as well, and these helmets seem to be of a higher quality manufacture suggesting these were for officers. Additionally, some examples may have been lined in green cloth as well as the same color of the outside cloth. The six-panel version of the first pattern helmets (First Pattern B) is the one mostly frequently encountered, and thus there were likely more of these produced than the First Pattern A models with four-panel construction.

At the top of both helmet patterns was a ventilator cap, and on each side the helmet was adorned with metal shields based on the insignia from the German Model 1935 "stahlhelm."

These were made of aluminum, non-ferrous metal or zinc, and were 4.2 cm high and 3.5 cm wide in the shape of a shield. These pierced the helmet, were attached with pegs, eyebolts or even screws. On the right side of the helmet was worn the national tri-color shield while on the left side was worn the eagle insignia shield of the Heer (army) in silver or gold for Kriegsmarine (navy), or an eagle in flight for the Luftwaffe (air force). The Waffen SS troops wore the SS runes on the left side with the "party" shield with the black swastika within a white circle on a red background on the right hand side.

In addition, the Kriegsmarine also wore a single gold eagle at the front of the helmet for marines and officers, and these were usually worn on white helmets.

The "Felt" Sun Helmet

The other main sun helmet pattern produced by Germany during the Second World War was the pattern that is sometimes called the Second Pattern or Second Model, or simply the "Felt" sun helmet. These helmets were likely used throughout Southern Europe, but evidence suggests that few (if any) ever made it to the North African campaign. While a sturdy helmet the darker color may not have been the best choice for use in the hot sun.

These helmets were made of pressed felt, which resulted in a seamless dark green-brown color. The underside of the helmet was a lighter green. This pattern

An example of the second or "Felt" pattern sun helmet.

The liner of the second or "Felt" pattern sun helmet.

featured a liner system that was made of a pressed felt shell with a leather headband, and this entire band then "floated" in the dome of the helmet via six spacers. Throughout this shell were a series of large round holes to provide airflow to the ventilator cap at the top of the dome. The same basic chinstrap system was used with these felt helmets.

This pattern helmet was likely produced in very large numbers, and many examples were never issued during the war. Throughout the 1950s and 1960s these surplus helmets were often sold with a "Made in Germany" stamp on the leather band.

Other sun helmets of the Third Reich

Throughout the African campaign, and during the occupation of Southern Europe the Germans used a seemingly endless variety of captured foreign helmets. The German naval forces in Greece apparently captured a large number of helmets, which were used throughout the occupation of the Greek islands. These were often adorned with naval insignia and in some cases even the black cap bands.

French Model 1931 helmets were likely captured in large stockpiles, while captured English Wolseleys and Indian pattern helmets were also used, and some surviving examples suggest that the German soldiers took to reusing the metal shields on these. These captured helmets were also used throughout the war by the German Polizei (Field Police) units. Additionally the Germans also used Italian Model 1928 pattern helmets throughout the war. In all these cases however collectors are advised to be cautious as surplus – as well as fake – metallic shields are readily available and have been used since the 1950s to "enhance" plain helmets!

An 88mm gun in action in the Western Desert with the Afrika Korps.

Chapter 4

Italian Helmets - Casco Coloniale

Italy was a late arrival on the colonial scene, and throughout the later 19th century suffered a number of significant setbacks in Africa. This did not stop the Italians from attempting to rebuild a colonial empire around the Mediterranean Ocean.

Throughout a lengthy war in Libya, which began in 1911 and continued through various colonial uprisings well into the 1930s, and the later invasion of Ethiopia in 1935 the Italians adopted and refined a suitable uniform for desert combat. So successful was this uniform the German Afrika Korps adopted a similar uniform – but utilized better tactics when fighting the enemy.

Model 1887 Colonial Pattern Helmet

From the first African campaigns in the 1880s the Italians were outfitted with tropical helmets. Among these early colonial pattern helmets were those likely of Anglo-Indian origin, including ones made of sola pith and cork. By the late 1880s the Italians had adopted their own pattern helmet, which was likely based on the British Colonial pattern Foreign Service Helmet. This helmet was the Model 1887 and was lightweight, worn without a puggaree – although some period photos and surviving examples suggest that some officers may have opted to add a cloth wrapping around the helmet, but this should not be considered the norm.

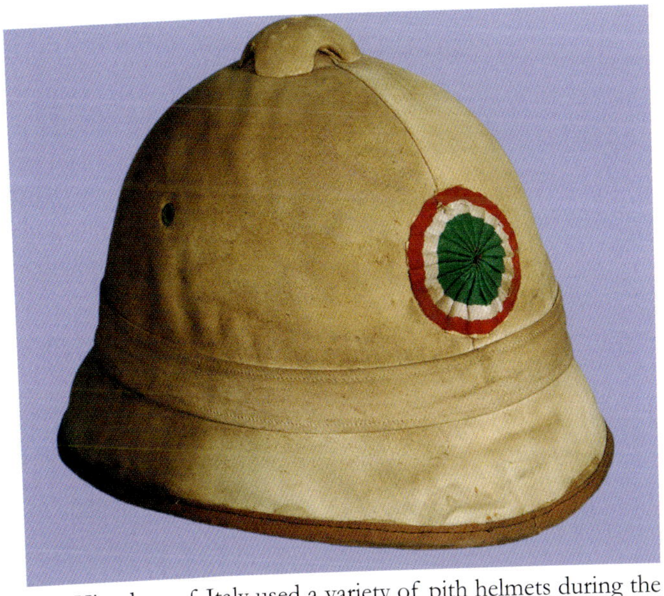

The Kingdom of Italy used a variety of pith helmets during the early days of their colonial empire building. These included British-made examples, as well as those made domestically by hat makers. This example is of the 1887 pattern, and may have been produced sometime in the 1880s or 1890s. This model remained in use until the introduction of the Model 1928 helmet.

The front view of the Model 1887 helmet with the Italian cockade and a rare badge, which may have been an insignia from a Carabinieri bandolier.

The helmets were covered in either white or khaki cloth, and for service overseas a helmet cover was also used. The helmets were of a simple pattern with a high dome covered in cloth, along with a flared out brim. There appears to have been a varying number of cloth panels used to cover the helmet, and these range from as few as two panels split evenly with the seams at the front and

An Italian soldier wearing the model 1887 helmet with cockade and feathers.

This pith helmet was likely made in India after the turn of the century, and is of the pattern most commonly referred to as the Model 1911 Sun Helmet. It does not feature the cockade, and most interesting is marked with the name of "LIEUT COM C.S.JACKSON.- H.M.S. 'VENUS'."Possibly this helmet was acquired by a British sailor in World War I or World War II. No other information is available.

The interior of a model 1911 helmet.

A 3/4 side view of the M1887 pattern.

back of the helmet, to as many as six panels along the lines of the British and French colonial pattern helmets, while the cloth around the brim could range from a single piece of fabric up to as many as six panels. At the top of the helmet was a screwed on ventilator, and Italian helmets are usually identifiable by the side vent holes, which have either steel or brass grommets. The Model 1887 helmet typically featured a tri-colored cockade of red, white and green (with red being outermost) on the front, along with an arm of service superimposed. The helmet badges were the same as those worn by the metropolitan army.

The liner consisted of a very thin leather band, which was supported by a number of cork bumpers, and the chin strap was sewn to the side of the helmet for enlisted men's models, while officer's models often featured brass hooks. Some examples featured a side pocket sewn on the right hand side for the wearing of feathers by Bersaglieri and other elite units.

The Model 1887 helmet pattern was refined in the early years of the 20th century, and this variation has come to be known as the Model 1911, likely due to its use in the invasion of Ottoman-controlled Libya during the Turko-Italian War. This helmet saw service throughout the First World War and into the 1920s. The helmets still featured the basic tricolor cockade on the front, along with a badge device for the arm of service that was either brass or silver, and these incorporated the regimental number. In 1895 the infantry badge had changed from a five-pointed star bearing a cross in the center to one with crossed rifles with a crown above and a disc in the center with the regiment's number.

Model 1928 Helmet for a New Roman Legion

The colonial pattern helmet remained in use well into the 20th century. It was finally taken out of service by the end of the Libyan uprising in 1931. The new helmet pattern

Soldier with M28 helmet and rifle (Courtesy collection of Paolo Marzetti).

was the Model 1928, but it was not widely issued throughout the army until the invasion of Ethiopia in 1935. This helmet was probably based on the British Wolseley pattern sun helmet and featured a similar shape and profile.

The helmet was made of cork and featured a rounded dome, and was covered in a thin cloth cover – usually

The feather pocket which was a feature of the M28 helmet.

cotton and khaki in color – and of six even panels. The brim of the helmet was pointed at the front and just slightly curled up at the back, and it also featured six panels of cloth covering. As it was still common for Bersaglieri units to wear their feathers on the right side, the Model 1928 helmets also feature a side pocket on the right hand side, but ironically all Model 1928 helmets feature a pocket for this option. Whether it was intended for the other units to eventually wear feathers with their sun helmets is not known, but this small pocket is a unique and most distinguishing feature of the Italian helmets of the World War II era.

Modeled after the British Wolseley pattern Foreign Service Helmet, the Italian Model 1928 was first introduced in 1935 and remained in service until the end of World War II. This example features the insignia of the 10th Granatieri (Grenadier) Division, which served with the 65th GRANATIERI DI SAVOIA in North Africa at the start of World War II.

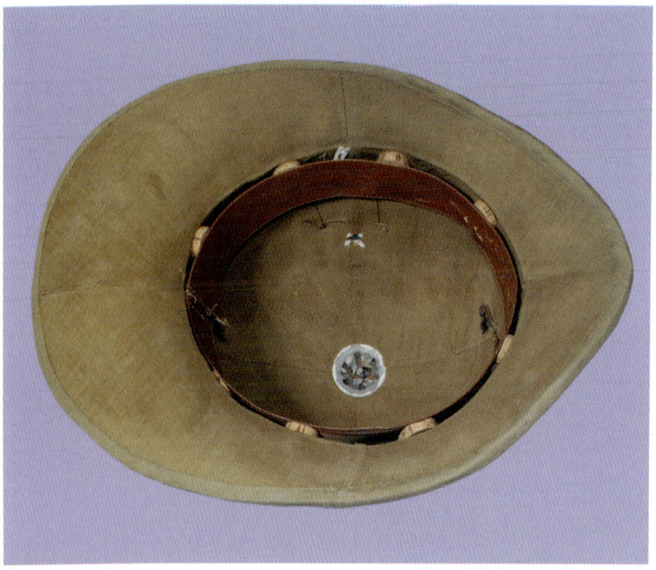

The interior of an M28 helmet showing the usual cork spacers which provide ventilation between the headband and the helmet shell.

An M28 helmet with goggles (Courtesy collection of Paolo Marzetti).

Also interesting is that the helmets featured a tan painted metallic ventilator cap at the top of the helmet, as opposed to the traditional cloth covered ventilator cap. These are usually found to be of non-ferrous metals including aluminum, and it is worth noting that the base into which these screw featured a mesh screen, possibly to keep insects or other objects from entering the helmet. Likewise, the Italian helmets retain the side vents, and instead of small holes with equally small grommets, these are larger – slightly larger than the size of a pencil – and these too feature a mesh screen.

As with the previous pattern the liner consisted of a very thin leather band, and the inside lining of the Model 1928 was either green or khaki. The materials available to the manufacturer likely determined the color.

An Italian soldier artillery-spotting and wearing the M28 helmet (Courtesy collection of Paolo Marzetti).

The Italian Model 1928 was never issued with any form of puggaree, however any Italian soldier lucky enough to obtain desert goggles was likely to wear these on the helmet when not used to cover the eyes.

As with the previous models, units of the Italian Army typically wore the 1928 pattern helmet with an Italian tri-colored cockade on the front, and these vary greatly in size. In addition on the front of the Model 1928 helmet was worn the brass arm of service badge. These also varied in size and included the regimental number. Period photos, as well as surviving examples suggest that the embroidered insignia was used to a lesser extent with some Model 1928 helmets, this must be considered the exception to the rule – and these examples may be from lower level company grade officers.

Some sources also suggest that as the war in the desert progressed that helmets, which were issued without badge and cockade, may have had the unit identification "stenciled" on – as was the common practice with the Model 1933 steel helmet. However, while it is possible that stencils were used it is more likely that these were printed or sketched on the sun helmets. This practice was not authorized but it was tolerated, and for these reasons it was unlikely to have been seen on the helmets used by the Blackshirts, such as the MVSN (Milizia Volontaria per la Sicurezza Nazionale) or CCNN (Camicie Nere).

The Model 1928 was originally designed as headgear for foreign service, but did see use in Southern Italy, Sicily and elsewhere. It was popular with various Italian Fascist paramilitary units, which also used the Model 1928 helmet. These units typically only wore an arm of service badge or some other sort of insignia, but usually not with the tricolor cockade. Thus Italian helmets are frequently encountered with unique fascist badges on the front, but it is up to the collector to determine when these were added.

This Italian Model 1928 sun helmet features the badge and feathers of the 8th Bersaglieri. In February 1941 the 8th Bersaglieri regiment landed at Tripoli and were incorporated into the Ariete Armoured Division. The XX Motorised Corps working in support of the 132nd Tank regiment were the most active and well known heavy armored unit in the Italian Army.

Another view of the 8th Bersaglieri helmet showing the side ventilation holes and grommets fitted with mesh screens.

A habit among some Italian soldiers was to attach captured Allied badges and medals, and this soldier above appeared to have covered his helmet with these "talismans". The strap over the crown of his helmet would indicate that this helmet is a captured British Bombay Bowler (Courtesy collection of Paolo Marzetti).

Italian Use of Anglo-Indian Helmets

Given the class differences in the Italian Army in the 1930s it was not common for many officers, especially those of field grade rank or higher to use the simple Model 1928 helmet. While it is a well-made helmet, it was regarded as headgear for the ranks, but many officers opted for sun helmets rather than the traditional visor cap or field caps.

An example of a straw M28 helmet, most likeley a result of wartime shortage of preferred material such as cork (Courtesy collection of Paolo Marzetti).

A Bombay Bowler style helmet featuring the badge of a light infantry regiment (Courtesy collection of Paolo Marzetti).

Thus it was common for Italian officers from the late 1930s to use private purchase Anglo-Indian helmets. These have typically been identified as being "Aden Pattern" or "India Pattern" helmets, and these are believed to be the Cawnpore and Bombay Bowler helmets respectively – though where the nomenclature of "Aden" or "India" comes from isn't quite so clear.

What is clear is that these helmets of both patterns are frequently seen in period photos. Especially popular with higher ranking Italian officers were the quilted "Aden" pattern helmets, and the khaki color of the late era Cawnpores did match closely with the Italian Model 1929 tropical jacket known as the Sahariana. These typically featured embroidered insignia (for a visor cap) rather than the metallic arm of service devices – although the latter is not impossible.

It is generally believed that the Anglo-Indian pattern helmets were likely purchased prior to the outbreak of hostilities, although it is possible that some helmets were captured during some of the reversals suffered by the British to Rommel's Afrika Korps which the Italians fought along side of from 1940 through 1943.

Chapter 5

Belgian, Portuguese and Spanish

Despite the reduced power that the states of Belgium, Portugal and Spain wielded in Europe in the latter half of the 19th century these three nations mustered colonial armies that defended each of their respective vast empires. Each of these nations also tended to look to the other major colonial powers when it came to uniforms, and notably, headgear.

Belgium

In the case of Belgium the empire wasn't so much for the nation rather than the personal fiefdom of Léopold II. He was the Congo Free State's first ruler, and this territory's peculiar constitutional status meant that it was unable to call on the service of the Belgian army. Instead it relied on mercenaries to make up the "Force Publique," the private armed forces of the Congo Free State. From the 1880s to the end of the 19th century the white officers of the Force Publique relied on a uniform that seemed far from practical for the tropical environment. This included a blue double-breasted frock coat for officers and a single-breasted blue tunic for NCOs. However, the Belgians had learned early on that the proper headgear was necessary for the tropical regions of the Congo, and a white sun helmet was issued to all white troops. These early helmets were originally of British and later French manufacture and featured a badge that consisted of a wreathed and crowned shield bearing a lion rampant (in gilt for officers, silver for NCOs and bronze for other ranks).

Belgian troops on campaign in the Congo circa 1918.

An example of the Belgian white colonial helmet. This one was worn in the Boxer Rebellion of 1900-1901.

In 1908 Belgium took over the administration of the Congo from the Free State, but despite this the Force Publique remained the primary armed force in the region. In the years leading up to World War I the 19th century pattern helmets were replaced by a helmet that closely resembled the British Wolseley helmet. The same type of badge was worn, but now with a tricolor Belgian cockade

The pattern of helmet, not unlike the British Wolseley, that superseded the earlier 1880s pattern sun helmet for the Belgian forces.

below. This type of helmet was used throughout World War I, with the badge being dropped following World War II – when only the cockade remained as standard until the Congo Crisis of the early 1960s.

Portugal

Tiny Portugal was the first of the European powers to have colonies in Africa, and ironically was also among the last. Despite this fact the nation's empire expanded little – in part due to its territorial ambitions being at direct odds with those of Great Britain. The result would have lasting repercussions that would eventually bring down the Portuguese government in the revolution of 1910, more than half a century before the end of Portuguese rule in Africa in 1975.

Used during the First World War by the Portuguese army in its African colonies, these helmets were made of a single piece of pressed felt.

During the second half of the 19th century Portugal was still recovering from the devastation of the Napoleonic Wars as well as the loss of its colonial empire in South America. This spurred renewed interest in Africa, where its army was primarily dressed in the same basic uniform patterns as worn in Europe. Only in the 1880s did Portugal begin to adapt and utilize a more appropriate uniform for the tropical climates of Angola and Mozambique. This included felt hats with white "Havelock" neck curtains, although period photos suggest some officers likely obtained white sun helmets – probably of Anglo-Indian origin.

It was not until 1901, when the overseas forces became a separate Colonial Army (along the same lines as those introduced by France), that sun helmets were more typical of the Portuguese forces in Africa. While many officers and NCOs continued to wear slouch hats – turned up on the left, as was the style of the Germans at the time – many white troops of the Colonial Army also wore a white sun helmet. These were likely of Anglo-Indian or French design and manufacture, and were worn with a short, removable brass spike. The Colonial Army helmets were often adorned with the headdress cockade and

Portugese troops embarking in Angola during World War I.

insignia. This included a blue and white cockade (mid-blue with a broad white rim) along with an eight-pointed star with the unit device in the center. While in the field

Throughout its long history as a colonial power the Portuguese used a variety of patterns of sun helmets, this is likely the final version. It is made of cork, and has the basic shape of the French Model 1931 helmet.

leather chinstraps were usually worn, but brass chinchains were part of the "dress uniform."

Beginning in 1911 the Metropolitan Army in Africa also adopted a sun helmet. These included foreign purchased helmets, including the British Wolseley and Cawnpore helmets. These were also worn with a small spike. On the front of the helmet was a red and green cockade for the Metropolitan Army, along with a gilt or bronzed arm-of-service device.

During the First World War the Portuguese army in Africa also began to use rather simple pressed-felt helmets. These were usually made of a single piece and were pressed into the standard colonial pattern shape. On the front was worn a nine-pointed star with the unit device in the center, as well as the short brass spike on top.

As the war progressed the badges were removed and replaced with a simple red (outer band) and green cockade on the front. Some units also began to wear the regimental numbers in brass below the cockade.

And while Portugal was to remain a force in Africa for more than another half-century, sun helmets were used infrequently in the decades to follow. This has made the surviving examples extremely hard to find.

Spain

The Spanish Empire had once been mighty and powerful, but by the age of the sun helmet, beginning in the middle of the 19th century, the empire was in decline. Spain was reduced to a few scattered colonies in the Caribbean, a handful of lands in Africa and the rest of the empire in the distant Philippines. Throughout much of the 19th century Spain, like many other European powers, relied on the same basic army uniforms used in Europe for the colonies – this despite centuries in these often tropical lands.

The Spanish called their sun helmet either a "capacete" or a "salacot," with the latter name being of Filipino origin. This is because Spanish troops in the Philippines were the first to have a regulation helmet, beginning in 1871.

This helmet was of Spanish design with a comb on the top, and was open at the front for ventilation – much like the Anglo-Indian airpipe helmet of roughly the same era. Whether these helmets were developed independently of one another isn't clear. However, the only known surviving example is at the Museo del Ejército in Toledo, Spain. This example appears to be made of a papier-mâché like material.

A Spanish Model 1894 sun helmet, which was authorized for wear with both dress and field service by a Royal Order of June 8, 1894. These helmets were likely copies of the French M1886 helmet (Courtesy collection of Bill Combs).

A later 1870s photo of a Spanish officer of the 7th. Infantry Regiment "Manila". He holds a Model 1878 salacot with a cockade and regimental number "7" on the side (Courtesy collection of Bill Combs).

This helmet pattern was short lived and was replaced in 1878 with copies of the contemporary British colonial pattern helmet. These cork helmets were covered in white drill, and bore a red and yellow cockade and brass unit number on the left side.

Another new pattern helmet, also of British style, was adopted in 1887. This particular pattern had a brass or nickel branch of service badge on the front while a spike was added for full dress. A surviving example, which was

The Model 1894 Spanish sun helmet was adorned with a brass or nickel front plate showing the royal arms of Spain (Courtesy collection of Bill Combs).

A Spanish made copy of the British Wolseley pattern sun helmet. It featured the same basic shape and design with some minor variations, such as the side ventilation hole with grommet. (Courtesy collection of Bill Combs).

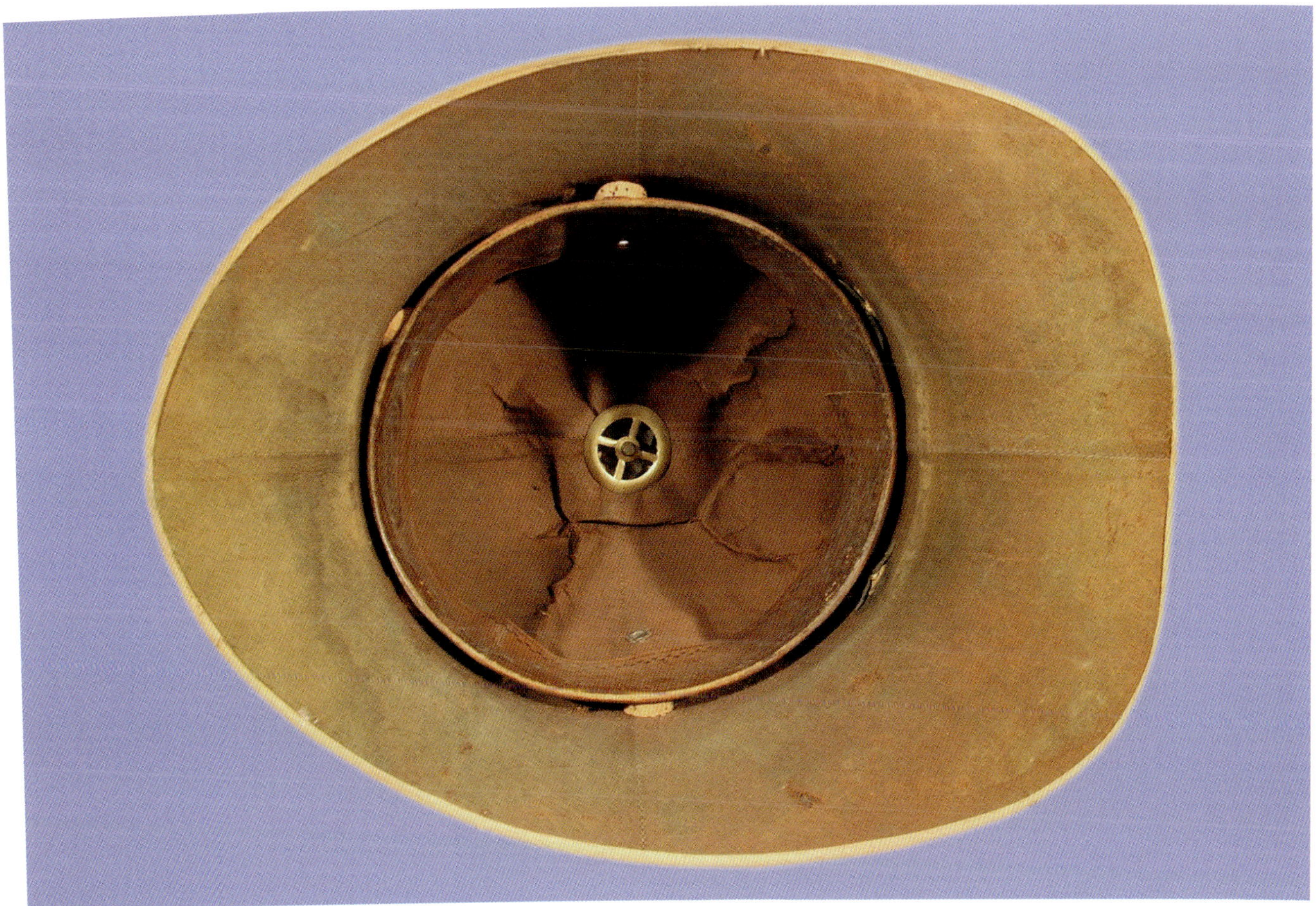

The liner of the Spanish copy of the British Wolseley pattern sun helmet (Courtesy collection of Bill Combs).

captured during the later Spanish American War, appears to have been made in Hong Kong. It is likely that this may have been the primary source for all the Spanish sun helmets issued in the Philippines.

The Governor's Bodyguard of the Philippine Islands also wore a white sun helmet of the same pattern as above, but it was primarily used for more formal occasions. It was fitted with ornamental brass spike and had a red leather band passed around the body of the helmet. It featured a badge on the front of gilt brass that included the royal crown of Spain, as well as the arm of service.

In the 1890s the Spanish adopted the colonial uniforms that are widely recognized and have come to be known as the "pajamas" pattern. These were made of rayadillo and were used throughout the remaining Spanish Empire. A variety of hats were used, including infantry pillbox caps as well as the more familiar straw hats that American forces in the Spanish-American War typically encountered.

A 1914 photo of two officers dressed in the Model 1914 khaki uniform. The gentleman on the viewer's left wears a Wolseley style helmet while his friend on the right has a Bartolini helmet (Courtesy collection of Bill Combs).

An officer of the 40th. Infantry Regiment "Covadonga" wearing the uniquely Spanish striped rayadillo uniform. In this case the regulation pattern used between 1911 and 1914. He holds a khaki Wolseley pattern helmet (Courtesy collection of Bill Combs).

In the years prior to the war, many Spanish officers – especially those of affluence – opted for sun helmets. These were also private purchase and may have included British and even French made helmets. These were typically worn without the puggaree cloth wrapping, but occasionally with a leather band around the side to denote unit type. On the front of the helmet was worn a wreathed royal coat-of-arms, or at other times the red and yellow cockade that was worn on the other Spanish headgear of the era.

Additionally it is worth noting that regular Spanish Army troops stationed in Puerto Rico were authorized to wear a pith helmet for both dress and field service by a Royal Order of June 8, 1894. The order called for these to be of British pattern but photos show that these were more likely copies of the French Model 1886 helmet. These had a brass or nickel front plate showing the royal arms of Spain. It is not entirely clear where these helmets were made, and whether these were produced in Spain or imported from France. While the regular Spanish army troops in Puerto Rico wore the sun helmets, the volunteer forces fighting for Spain continued to wear straw hats.

The American forces fighting against the Spanish in Cuba likely encountered few if any soldiers wearing sun helmets. This is because sun helmets were not actually authorized for troops in Cuba, where the straw hat or "sombrero jipijapa" remained the crowning glory of the Spanish soldier.

After the loss of the overseas colonies in 1898, Spanish troops saw extensive action in North Africa. In 1909 a large number of British Wolseley pattern helmets

A 1914 photo of a young Light Infantryman wearing the Bartolini helmet. (Courtesy collection of Bill Combs).

were purchased from the British at Gibraltar for use by the troops fighting in Morocco. These were reported to have been white with a separate khaki cotton cover. By 1912 these British helmets were being replaced with Spanish made copies. In 1913, 30,000 French pattern helmets were also purchased and became known as the "Bartolini" pattern, after the name of one of the French manufacturers. These were likely worn plain without any insignia or other device. Both the Wolseley and Bartonlini pattern helmets continued in simultaneous service until being gradually replaced after 1917 by the khaki canvas brimmed "Daisy May" style hat known as the "Americano."

Chapter 6

American Sun Helmets - From Cork to Pressed Fiber

One little known facet of the history of military sun helmets is that it is entirely possible that among the first general officers to wear such headgear outside of India – following the Indian Mutiny from 1857 to 1859 – one may not have been British. While the British Army was quick to adopt the Indian pattern headgear, it may have been an American cousin who used the helmet at one of the most important battles of the American Civil War, albeit the losing commander!

While there were no photographs taken of Brigadier General Irvin McDowell at the Battle of Bull Run on July 21, 1861, accounts suggest he was wearing an "outlandish" helmet that resembled a basket on his head. It has been suggested that he could have been wearing a cavalry helmet, but considering that McDowell had traveled throughout India in the decade before the American Civil War it is entirely possible that he acquired some form of Anglo-Indian sun helmet. Considering the outcome of the battle it is probably best that the exact type of headgear is lost to time.

An illustration by W.T. Clark of the 4th New Hampshire Infantry as they prepared their fortifications at Hilton Head. Note that it appears that these Union soldiers are wearing an early pattern sun helmet, but little is known about this headgear.

The widely produced Model 1887 sun helmet. Despite rumors that these helmets were British made, these helmets were produced in large numbers by Horstmann Brothers and Company of Philadelphia, a well-known dealer and manufacturer of military uniforms.

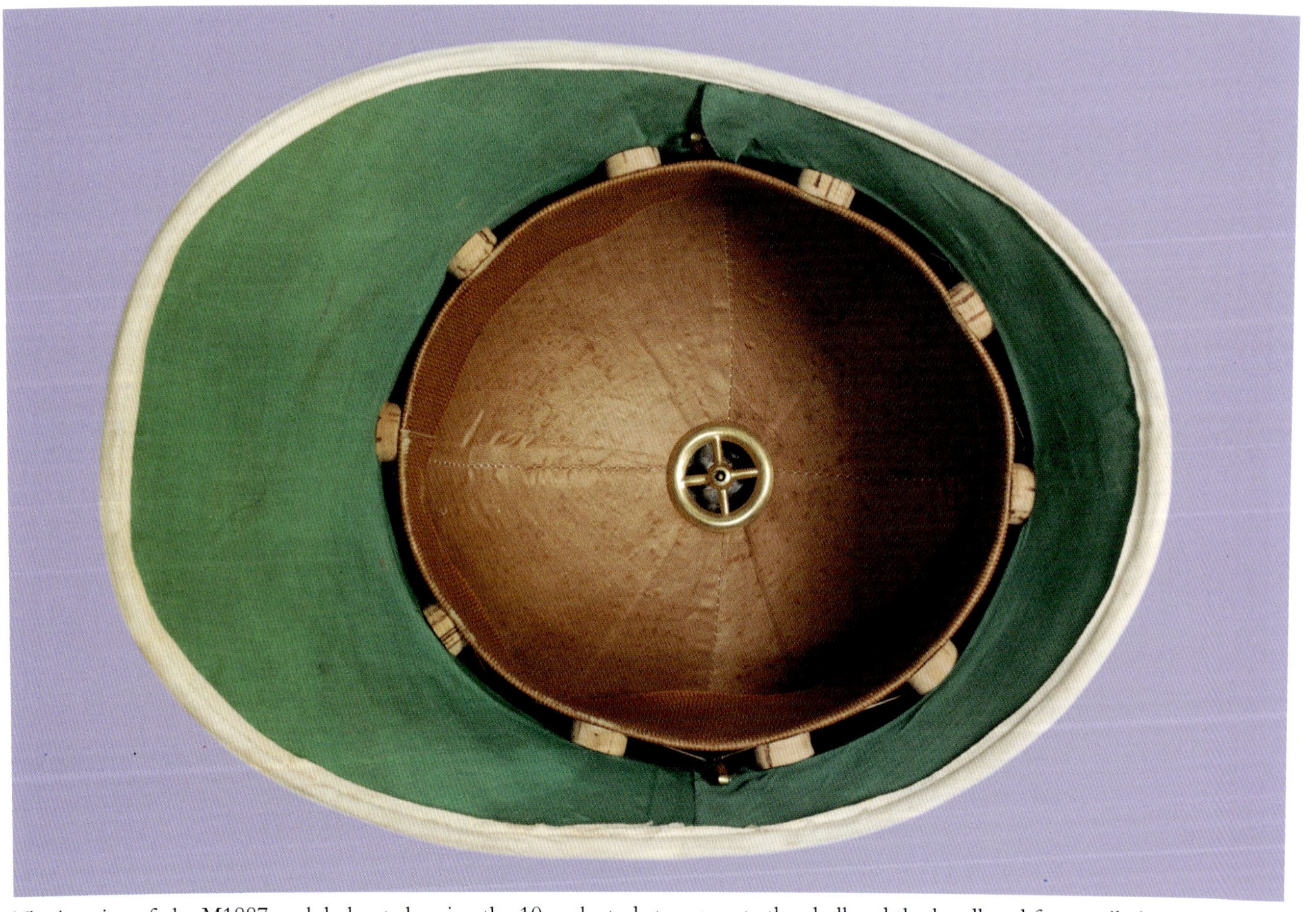

The interior of the M1887 cork helmet showing the 10 cork studs to separate the shell and the headband for ventilation.

Some of the contemporaries of McDowell's may have noticed the helmet. Sources, including Philip Haythornthwaite's 1976 color uniform book "Uniforms of the Civil War: 1861-65" even suggest that some Union officers may have used sun helmets for protection from the sun, and that "The 4th New Hampshire Infantry wore, for a period, a 'pith helmet' or sun helmet, designed to protect the wearer from the heat of the sun. Such helmets were not uncommon (particularly in the earlier part of the war)."

However it was not until the early 1870s that the United States Army looked to reintroduce a summer uniform of lighter material, along with a new type of summer headgear.

During the same period the Americans were clearly inspired by the success of the German armies in the Franco-Prussian War, and adopted a dress helmet that very closely resembled the famous Pickelhaube (spiked helmet). This helmet of black felt was worn with a spike, as well as horsehair plumes, featured a helmet plate on the front for arm of service and even included brass chinchains for officers.

It was practical as part of a dress uniform but was entirely inappropriate for the role the United States Army played in the Indian Wars on the Frontier. Likewise, the helmet was far from ideal for the summer heat. For summer time headgear the Americans considered a number of options including the wearing of straw hats, but eventually the American military looked to the British for a summer helmet.

Cork Helmets of the 19th century

In the early 1870s, even before the British formerly adopted the Foreign Service Helmet, the United States Quartermaster at the time, Montgomery C. Meigs, contacted the British government to obtain cloth-covered cork helmets. This may have lead to one of the greatest misconceptions about the American sun helmets – that is that the helmets were British made. In fact there is no proof that any of the helmets issued to the United States Army in the latter years of the 19th century were of British manufacture.

These first helmets were approved by the United States Army as the summer helmet for infantry and cavalry units, and over the next three years were issued experimentally to troops of the 9th Cavalry serving in Texas, and to troops at the Artillery School in Fort Monroe, Virginia. Horstmann Brothers and Company of Philadelphia likely produced the original Model 1880 pattern helmets domestically, again despite the long accepted belief that the early American sun helmets were made in England.

On May 5, 1880 the Secretary of War approved the specifications for summer helmets for enlisted personnel as follows:

Shape and weight – To be in shape according to standard sample and to weigh about seven and one-forth (7 1/4) ounces when finished; reasonable variations (from this weight) due to sizes to be allowed.

Material, etc. – The shell to be composed of two thicknesses of the best quality of cork, laminated or scarf-seamed, and securely cemented together with shellac. The linings to be firmly shellacked to the inside of the shell; that for the dome to be of slate-colored drilling, and that for the visor or shade to be of emerald-green merino or cashmere. Sweat-leather to be on frame or hoop as in sample, well separated from the shell (for ventilation) by ten (10) small cork studs securely fastened; sweat to be about one and three-eights (1 3/8) inches deep, and to be provided with a drawing string. Outside covering to be of the best quality of bleached cotton drilling, in four (4) sections, welt seamed and secured to the shell with shellac. Band of same material, about three-fourths (3/4) of an inch deep. Edge to be bound with stout bleached stay binding. Adjustable ventilator at top as in sample. Chin-strap of white enameled leather, and brass hooks for same, as in sample.

An M1887 cork helmet of the 4th Artillery. Many American units used this style from 1880 to 1905. While these white helmets were never officially issued with brass accoutrements, many officers applied dress spikes and front plates. This practice was banned by 1905 when the army redesigned their uniforms.

A Model 1889 cork helmet covered in tan cloth, as opposed to a white cloth helmet that was dyed tan.

About 6,000 of the Model 1880 helmet had been produced within a year. This first style of helmets, which had a shorter front and rear bill than the British Colonial pattern helmets of the same era, were as the Quartermaster General specified constructed of cork and covered in white cloth. These helmets featured four seams (rather than the six seams of its British cousin).

This helmet was modified to coincide with the new Model 1887 uniform, and this has lead to some confusion. Some helmets are referred to as the Model 1887, but the changes in the helmet designs really refer to the Model 1889. This later version featured a steeper front bill and a much larger rear cap. The basic shape and construction remained the same. However, as the British discovered, white sun helmets may look good on parade but the color proved impractical in the field, and this was changed in the 1882 to a khaki color, to make the wearer less visible to enemy fire.

As existing stocks included white helmets, these remained in use with the American military, including Army and USMC units, through 1904, and sources including Gordon Chappell's "Summer Helmets of the U.S. Army, 1875-1910" suggest that the helmets were worn until 1909. The various colonial pattern helmets saw service in the American Indian Wars, and in the Spanish-American War.

Nevertheless, the helmet was as unsuccessful to the Americans as it had been a success to the British. The American enlisted men hated the helmet, and most helmets were discarded quite quickly. There is however, ample evidence the helmet was better liked at least by officers in Cuba, and later in the Philippines. It could be that these helmets for the officers were of better quality – given that officers were required to purchase their equipment. Then again, it could be that given that all the equipment could be expensive the sun helmets were used because the officer felt it better to use them than have them go to waste.

There are numerous surviving examples of higher quality helmets, and it is believed that some officers may have had at least two sun helmets with one being for normal field wear and the other saved for dress occasions. As the United States Army never adopted helmet covers this seems like a reasonable assumption. Additionally, at no time did the Americans ever authorize the use of a puggaree with their helmets for either enlisted men or officers, and it seems unlikely that helmets were ever used with any.

The American summer helmets were authorized for wear in the field as well as with the summer dress uniforms in place of the Prussian looking black felt dress helmet that was used at the time. But interestingly the white summer helmet, as well as the khaki offshoot, did not become an authorized dress helmet until General order 197 was published on December 31, 1904. However, the summer cork helmet was never intended to

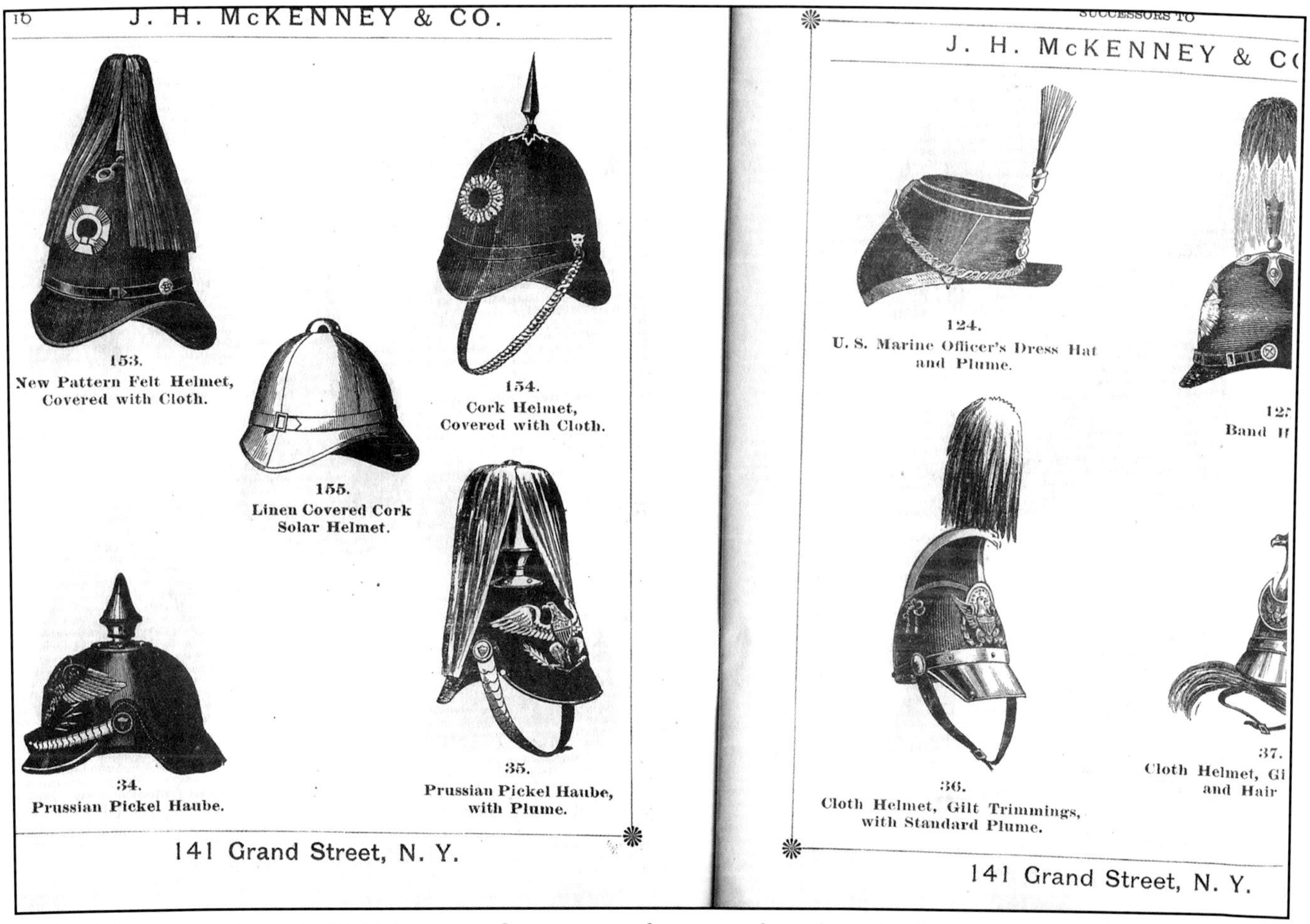

Ridabock & Company, formerly J.H. McKenney & Co., were manufacturers and suppliers of military goods, and white summer helmets were an item available for private purchase.

Privates Jones and Cherneck wearing the 1889 cork helmet.

be used with insignia or trimmings, but despite this fact it seemed that it was not uncommon for officers to transfer the eagle, oak leaf base, side buttons and cords from the black helmet.

This practice was officially banned in 1887, but numerous examples suggest that it remained in practice until the helmets were taken out of service. Likewise, because the helmets and accoutrements were sold as surplus in the Bannerman catalogs for many years, it seems likely that many examples today were "pieced" together by collectors over the years. Therefore trying to determine when a helmet was modified can be especially tricky today.

Several companies manufactured the white and khaki cork helmets, and the largest suppliers were Horstmann Brothers & Company of Philadelphia, whose contract to make helmets began in June of 1880, and J. H. McKenney & Company of New York, later Ridabock & Company. Its original contract began in December of 1885. These were the largest suppliers of military goods for the United States Army. Additionally, other manufacturers included Raymond & Whitlock of New York, Henry V. Allien & Company of New York, J.H. Wilson of Philadelphia, Charles Lehman of New York and the Manhattan Supply Company of New York.

Militia and Other White Helmets

Several gray areas exist in the topic of American sun helmets. The first was that it was common for National Guard and militias to use the helmets. As many of these helmets were purchased specifically for these units from a variety of suppliers there is a massive number of helmet variations. While many units used the same helmets that were available for the regular army, the helmets made specifically for militias and National Guard troops were smaller with less of a flared brim. These still were of the traditional four-panel construction, and depending on the state unit these may have had unit numbers on the front.

American M1880 helmet with the National Guard number of 31. This is likely for a New York unit.

McKenney & Company, later Ridabock & Company, were major suppliers of these militia helmets and are believed to have provided the New York National Guard with its helmets from the 1880s into the early 20th century. And while it might seem odd that a northern state would require helmets, many of the New York forts were on water, where the humidity could make it quite warm in the summer months.

Another area of confusion is when exactly the first

This model 1880 helmet is the New York National Guard variation made by Ridabock & Company, formerly Baker and McKenney. This example features the side ventilation that was popular with state national guard helmets, but also used by the regular U.S. Army.

United States Marine Corps units wore white summer helmets. Photographic evidence is sketchy at best, but it is possible that the USMC may have been wearing sun helmets as early as 1879 but proof has yet to surface.

The final gray area of confusion is in regards to the countless helmets that show up on auction sites, are found in flea markets and even make their way to militaria shows. These helmets contain parts from United States Army sun helmets, and in some cases may even have been manufactured by the same makers. A common distinguishing feature is a skull and cross bones plate on the front – but these are not Hussar helmets. Instead these are helmets from various late 19th and early 20th century fraternal orders and lodge associations such as the Odd Fellows and Knights of Pythias.

When the helmets were taken out of service in the first decade of the 20th century many original United States Army sun helmets, along with various spare parts, were sold through the Bannerman catalog, and thus many collectors should be wary of parts helmets. A telltale sign is a helmet with mismatched parts, such as infantry buttons on a militia helmet or a plate that doesn't match the rest of the helmet.

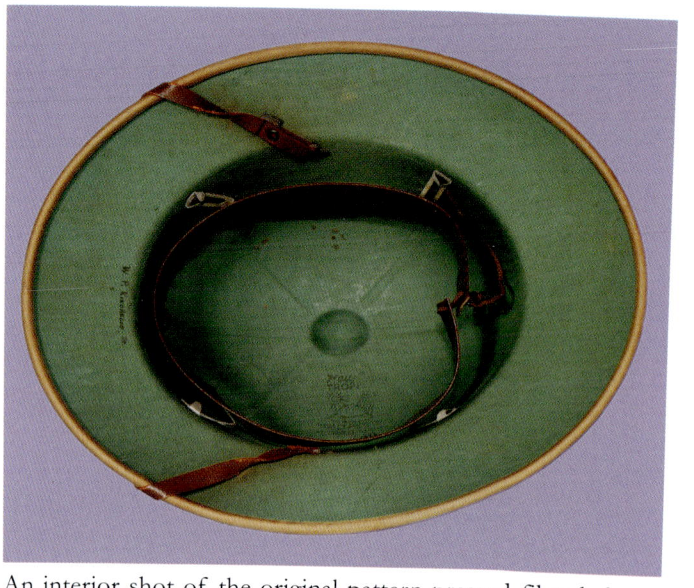

An interior shot of the original pattern pressed fiber helmets shows how the liner was secured to each side via two rivets, while the chinstrap was attached via a third rivet in the middle of the other two. This example features the early wartime leather chinstrap.

Fiber helmets From World War II to Vietnam

By the time the United States went "over there" and into the trenches of France during the First World War, the sun helmets had long since been removed from the American soldier's uniform. It was not until a generation later that the sun helmet reappeared.

A new sun helmet was introduced just before America's entry into World War II. Hawley Products Company likely designed the sun helmet, which unlike most American field gear seemingly has no specific model number! Hawley was the primary manufacturer of the helmets, which were made of the same basic materials as the M1 steel helmet liners, which called for a two-part fiber material to be formed into the basic shape, and then impregnated with a bonding agent to make it water proof. Examples were also made by the International Hat Company and both the wartime helmets made by Hawley and International Hat feature an ink stamp on the inside of the dome.

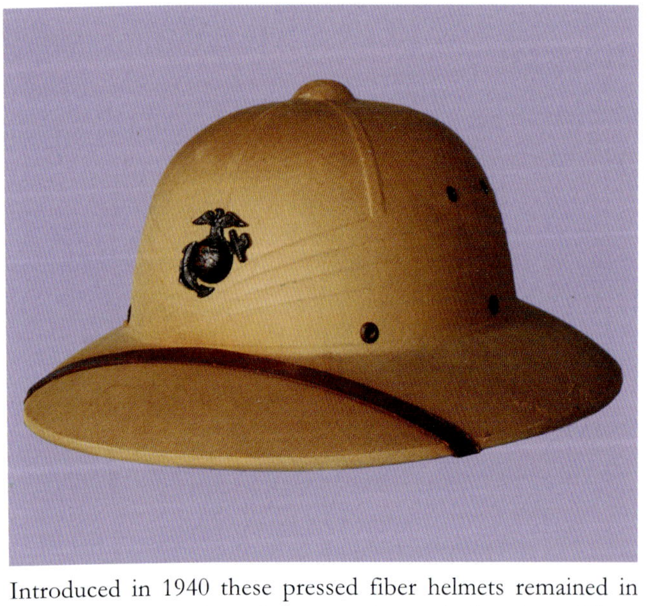

Introduced in 1940 these pressed fiber helmets remained in service for more than sixty years in tropical regions. These helmets featured the bronze EGA of the USMC.

American Naval officers in the Leyte Islands 1945 with the pressed fibre helmet with insignia.

An interesting example of a late war World War II American pressed fiber helmet, with a USN "officer cap badge" style insignia painted on the front. While the practice of writing on the helmets was discouraged it was fairly common for Americans to "graffiti" these helmets, especially later in the war. This example features the wartime leather chinstrap.

This examples features the lighter canvas covering, and these helmets were used by the US Army from World War II through the Vietnam War. It features the standard American Officer's cap insignia. It has a Hawley makers stamp and is dated to 1948.

The result was among the simplest sun helmets ever produced. The pressed fiber shell was then covered with a thin layer of cotton cloth, and painted green on the inside. It had the basic shape of a "safari" helmet, complete with a faux ventilator at the top and a faux wrapping of puggaree around the dome of the helmet – the latter ironic because previous American sun helmets never used a puggaree. For ventilation there were a number of vent holes with grommets on each side. The early helmets had three lower vent holes with two vents above, while post-wartime examples feature four lower ventilation holes.

The lining system was affixed to the two outer lower grommets, and consisted of a thin rubberized band and the chinstrap was attached to the center of the three grommets. The wartime chinstraps were of the same manufacture as the M1 steel helmet liner chinstraps. The later war lining changed slightly, with the liner band attached to the two inside (of four) grommets, and a hook for the chinstrap attached to the helmet band. The later examples also feature an elasticized strap, which was replaceable, unlike the early wartime example that was permanently attached

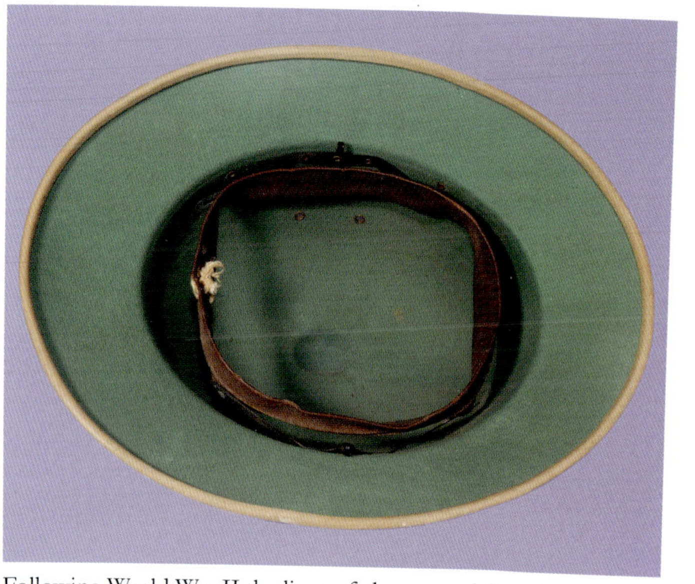

Following World War II the liner of the pressed fiber helmets was changed, with the chinstrap attached via hooks on part of the liner band on each side of the helmet.

A post WWII USN pressed fibre helmet.

The helmets were originally produced with a light khaki colored cloth that was the same color of the pre-war American uniforms. Both the United States Army and the United States Marine Corps used these helmets, and the basic cap badges were worn on the front through a hole with a grommet. Based on photos and surviving examples it seems that these helmets, which were designed for use in warm weather regions, were used mainly by army officers, who wore the gold officer's cap insignia on the front. The USMC appeared to use these helmets across all ranks, and these examples feature the Eagle-Globe-Anchor (EGA) cap insignia. In the early part of the Second World War the fiber sun helmets were used as the "training helmet" at various USMC bases, as steel helmets were not always readily available to the Corps. Additionally it is worth noting that the helmet was known in the Marine Corps as the "fiber helmet," and that the regulations following its issue in 1940 officially and simply refer to it as a "helmet, fiber."

Later in the war a darker version appeared in an olive drab color, and it was originally the United States Navy that used these helmets. However, surviving examples do suggest that the USN did use the earlier khaki versions, and in the post-war years and into the Vietnam War the OD version was used by all services.

The sun helmet predated the introduction of the M1 steel helmet, and remained in service throughout the Vietnam War and was used stateside in hot weather locations until the 1980s, also outlasting the steel helmet. In fact, while among the most under appreciated American helmets, it was the one that served the nation the longest.

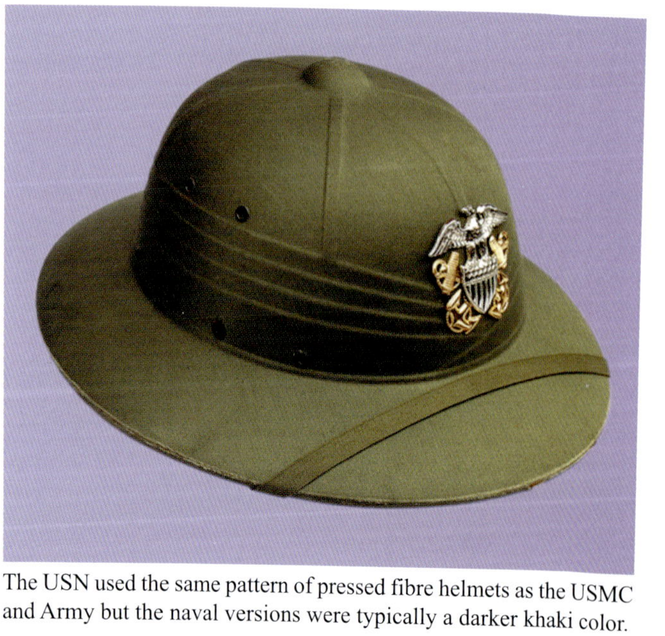

The USN used the same pattern of pressed fibre helmets as the USMC and Army but the naval versions were typically a darker khaki color.

An International Hat Company version of the pressed fiber sun helmet. While the same basic shape, a close inspection reveals that it has a greater number of "folds" in the faux puggaree than the Hawley made version.

The maker's label of the International Hat Company version of the pressed fiber sun helmet.

CHAPTER 7

JAPANESE - TROPICAL HELMETS OF THE NEW SAMURAI

During the latter half of the 1930s and throughout World War II the Japanese military issued an enormous number of sun or pith helmets for their forces, both in the Imperial Japanese Army and the Imperial Japanese Navy.

As there was fierce rivalry and competition between the two forces – which no doubt affected the nation's efforts to wage war successfully – much of the equipment was produced independently. For this reason it is extremely uncommon to see Japanese army pattern helmets used by the IJN, and for naval pattern helmets to be used by the army.

Japan's first efforts to obtain an "overseas" or colonial empire were actually just across the Sea of Japan to nearby Korea, which it officially annexed in 1910. This followed a hard fought war with rival Imperial Russia, which saw significant land fighting in Manchuria throughout 1904 and 1905. Some sources suggest that in the summer months the Russian officers may have worn sun helmets – possibly of German manufacture given that Germany's Kaiser Wilhelm II was a keen supporter of the Russian cause (and even an honorary Russian admiral).

It is unlikely that the Japanese army used any sun helmets at this point or even during the First World War, although obtaining private purchase items in Hong Kong or other European "outlets" in East Asia would not have been difficult. However, Japan did fight along side the British forces in the Pacific during World War I, and seized a number of Imperial German island colonies and in 1914 assisted with the capture of Tsingtao in China from the Imperial German East Asiatic forces – who had their own unique sun helmets (see Imperial German Sun Helmets).

Type 90 Style Sun Helmet

Whether the influence of British, German or even Russian sun helmets had any impact on Japanese military planners is not known. However by the mid 1930s Japan had designed its own simple sun helmet. This bowl-like helmet had the basic silhouette of the Type 90 steel helmet, and thus is known, at least among collectors, as

A studio photo of a Japanese soldier wearing the Type 90 style sun helmet, which was apparently based on the steel helmet (Courtesy collection of Jareth Holub).

A classic example of a Type 90 Style sun helmet. This was a later pattern, and was used from mid-to-late Word War II. This pattern had a dual purpose and could serve as a sun helmet or a steel helmet cover (Courtesy collection of Jareth Holub).

An early patternType 90 sun helmet with a top ventilation cap, which was likely based on European style vents. The use of these vents was discontinued in later helmets to simplify manufacture and to provide better cover in the tropical rain where water often leaked in through the ventilator cap (Courtesy collection of Jareth Holub).

the Type 90 style or pattern sun helmet.

It was constructed of six segments of khaki cloth cork with a strip of cloth fastened around the base of the helmet. The inside of the helmet was typically also covered in cotton, usually of a darker olive drab color. As with the steel helmets in use at the time this type was issued to all ranks and appeared with and without a cloth insignia on the front, although it was common to see the yellow Japanese star affixed to the front.

Unlike the steel helmet, which featured a traditional Japanese style cloth chinstrap that was tied around the wearer's chin, the sun helmets utilized a simple adjustable strap that was sewn into the helmet's inner fabric.

As with other Japanese equipment of the World War II era these helmets typically have two white cloth rectangular labels sewn into the interior of the helmet dome. These are divided into two or three sections with one tag being there to provide information including the soldier's name and other personal information. The other tag had the printed showa (year) date, depot and manufacturer's name. There were often additional inspector's marks in ovals located inside the helmet's interior.

Countless variations of the Type 90 style sun helmet exist, including those of a variety of different materials, some with lighter colored fabric on the outside, as well as various examples with ventilator caps on the top of the helmet or ventilation holes with grommets on the side.

A less common variation of the Type 90, this pattern has often been wrongly labeled "an officer pattern" because of its being worn by the Japanese commandant in the film *The Bridge on the River Quai* (Courtesy collection of Jareth Holub).

A variation of the Type 90 made of wicker with a cloth cover (Courtesy collection collection of Jareth Holub).

A rare example of a Type 90 sun helmet with factory applied waterproof coating. (Courtesy collection of Jareth Holub).

The interior of the above helmet which shows the exposed fiber material and lack of any fabric lining (Courtesy collection of Jareth Holub).

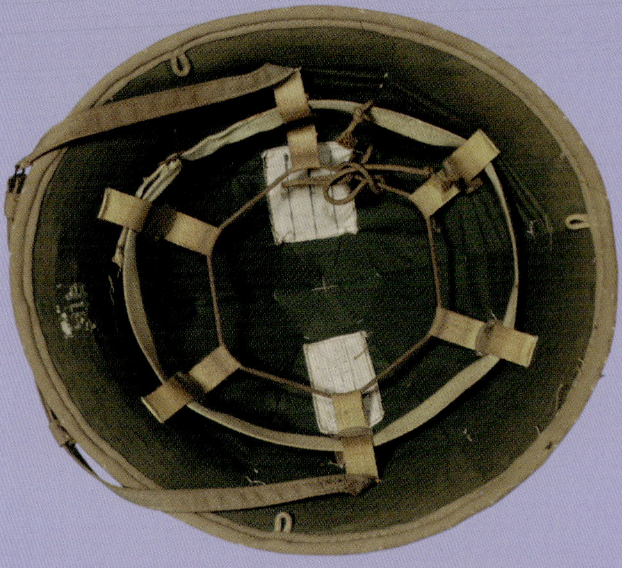

One of the more common, but no less unique, variations features three space string loops with one at each temple and one to the rear. These were used to help secure the entire sun helmet over the Type 90 steel helmet thus transforming the sun helmet into a cover! Based on surviving examples this seems to have been good in theory but was likely far less successful in practice (Courtesy collection of Jareth Holub).

Numerous variations of liner fabric can be seen with Japanese Type 90 style sun helmets, but the use of this blue lining was likely due to the material available to the manufacturer at the time of production (Courtesy collection of Jareth Holub).

A unique variation of a Type 90 style sun helmet. This example was made of wicker rather than cork and features no internal cloth lining, and inside the liner band is attached directly to the wicker. This is likely a later war-time economy measure (Courtesy collection of Jareth Holub).

While the most common examples were made of cork, wicker and straw examples are also encountered, as are those with common household fabrics used to produce the helmets, and these various ersatz materials are believed to be wartime economy measures. This example is made from re-cycled Kimono material (Courtesy collection of Jareth Holub).

The interior of the above helmet also features recycled cloth lining and the same cloth that is used for the exterior was also used for the cloth chinstrap (Courtesy collection of Jareth Holub).

Solar Safari Helmets of the Navy and Army

Given that the Imperial Japanese Navy had modeled itself after the British Royal Navy since the latter half of the 19th century it is not surprising that its officers looked to the West for their own types of sun helmets when the need arose.

However, it is worth noting that in the 1930s the rise of nationalism and imperialist ideas within Japan clashed greatly with those old notions. Thus the Japanese naval helmets are somewhat of a paradox, one that has not been fully understood by collectors in all the decades since the end of hostilities in 1945. In part for these reasons, there is no single pattern or more accurately "Type" of Japanese Navy or Naval Landing Force helmet.

The helmets used by the Imperial Japanese Navy do have a British influence to them, and are reminiscent of a "safari helmet," and among Japanese helmet collectors these are sometimes called the "European style" helmets. These are no doubt based on, and likely include examples of the British Cawnpore and Bombay Bowler helmets, as well as other civilian style Indian pattern helmets. These helmets were also widely used by the IJA. Surviving examples feature labels in the dome that indicate these were the Rabbit, Elephant & Swan brands, and were likely made in Japan in the late 1930s onward.

This helmet type was made in a number of patterns, and as with the Type 90 style sun helmet the variations are seemingly almost endless. The most common examples were cloth-covered cork helmets that were produced in a variety of colors including white, green and khaki. Interestingly the inside of the helmets were often exposed cork. The outside dome of the helmet was wrapped in a puggaree with the metal arm of service device affixed to the front (a brass army star or a gilt naval anchor and cherry blossom, which was similar to the

Close-up of Japanese Naval insignia (Courtesy collection of Jareth Holub).

Three Japanese Naval Landing Force personnel posing next to an armored car. The soldier to the far right is wearing a "safari" pattern sun helmet (Courtesy collection of Jareth Holub).

A Japanese Imperial Naval officer helmet. The safari style helmets were typically used by officers (Courtesy collection of Jareth Holub).

insignia on Naval Landing Force steel helmets), while some examples feature rank insignia on the side. These helmets featured ventilation eyelets with grommets on the sides.

As some of helmets were believed to be private purchase items, these further result in the endless variety of patterns. To further add confusion when the Japanese forces overran Hong Kong in late 1941 and Singapore in 1942, as well as the Dutch East Indies, large quantities of sun helmets – both military and civilian – were captured and widely issued to soldiers of the Japanese military. Furthermore, as Japan's wartime economy was already stretched to the limits, these helmets were widely adopted for use.

Unfortunately this has made it very easy for these helmets to be faked. Rather common helmets from India and Malaysia can be transformed into an NLF or Army officer's helmet with the simple addition of a wartime insignia. Thus collectors are cautioned when acquiring these types of helmets.

A close-up of the Japanese army star on the officer helmet (Courtesy collection of Jareth Holub).

An example of an army officer's "safari" style helmet with rank collar tab added to the side. This practise was uncommon but did occur in very hot areas where jackets may not have always been worn (Courtesy collection of Jareth Holub).

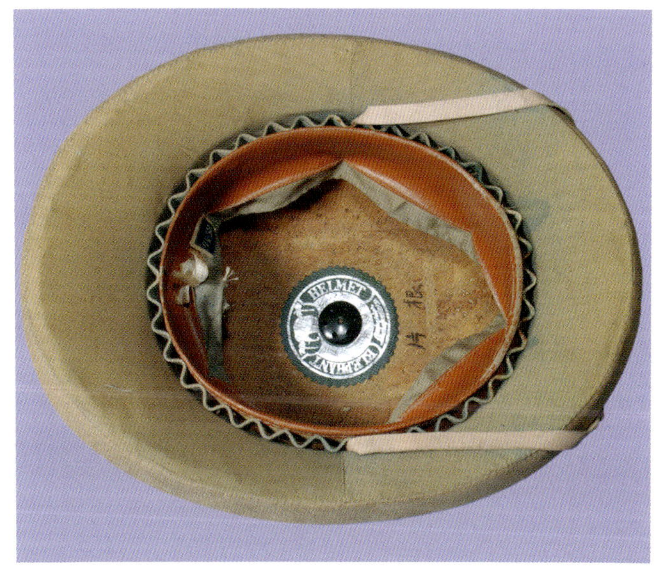

The liner of the above helmet including the manufacturer's label (Courtesy collection of Jareth Holub).

Additionally many sun helmets were found throughout the 1950s and 1960s in storage and these have occasionally flooded the collector's market as World War II issue helmets. While there is no doubt that these were manufactured during the 1937-1945 era these did not see service and thus some collectors do not consider them to be war era issue helmets.

Chapter 8

South American Sun Helmets - Argentina, Brazil, Peru and Uruguay

Throughout the 19th century and into much of the 20th century, practically all of the nations of South America looked to Europe as a model for military traditions, as well as uniform styles. While much of South America was originally part of the Spanish Empire, with the notable exception of Brazil (as well as French Guiana, Dutch Guiana and British Guiana), it is no surprise that the post-independence nations continued to follow the military traditions of their former masters. By the middle of the 19th century many nations followed the French and British models, and the various South American nations wore uniforms that resembled those in use by the powers in Europe. Even Brazil, which won its independence from Portugal, continued to be influenced by Portuguese and British military fashions.

The most significant change came with the Franco-Prussian War, and following the defeat of France in 1871 the nations of South America adopted uniforms that resembled those of the German Empire. Even with the German defeat in World War I, the spiked helmet (or Pickelhaube) remained in widespread use among these nations and is, in fact, still used by Chile today!

During the latter part of the 19th century these South American nations began to adopt summer uniforms, and with it sun helmets. While colonial pattern helmets were used to some extent the "safari" type helmets were more widely used. As the South American nations often lacked the industry to produce a range of equipment – including the sun helmets – these were widely imported, thus it is not uncommon to find American, British and other nations' sun helmets being used by the military forces of such countries as Colombia, Chile and Uruguay.

An Argentine model 1910 spiked helmet adapted from the German Pickelhaube.

Argentinean soldiers wearing a colonial pattern helmet based on the British design, circa 1900.

Argentina

At the dawn of the 20th century German advisors, along with those of Chile, helped establish the Escuela Superior de Guerra in Argentina. Founded in 1900, this military academy was part of the nation's plan to transform the Argentine army into a modern fighting force. This lead

The helmet badge of the Argentine Army model 1927 helmet is a smaller version of the "Picklehaube" (spiked helmet) style badge.

to the founding of a cavalry school in 1904 and a non-commissioned officers' school in 1908, and surprisingly a military aviation school was created in 1912.

With the arrival of the Germans, the Argentine uniforms changed from those of a French style to one far more German. Argentina officially adopted the spiked helmet (Pickelhaube) in 1910 for the army, as well as for the military schools.

The Argentine military and police forces also used a summer version of the spiked helmet, and this helmet was widely used until the 1920s. These helmets were of the basic colonial pattern shape with a tall dome and rounded front and rear visors. Khaki cloth of two panels covered the dome. The helmets were adorned with a large Argentine military badge in brass or steel on the front, along with a spike and rounded spike base at the top of the helmet. A leather chinstrap was worn around the outside of the helmet front, following the model of the Pickelhaube. The badge was unique in that it featured a sun rising from the top – and this may be one of the few cases where a sun actually appears on a sun helmet!

This pattern helmet remained in use until around 1925 when the spiked helmets were widely removed from service, although still used by the police into the 1930s.

A unique Model 1927 "casco Italiano" with the shorter brim to the front. This helmet is adorned with the Argentine state cockade and the badge of an airforce academy officer.

A front of the "casco Italiano" shows how this helmet was typically worn with the larger brim to the front.

A variation of the Model 1930 sun helmet; this example has no top ventilation, and instead, has only two forward vents. It also is a larger helmet, and has a front and rear brim of equal size. The high quality cap insignia suggests it was a private purchase officer's helmet.

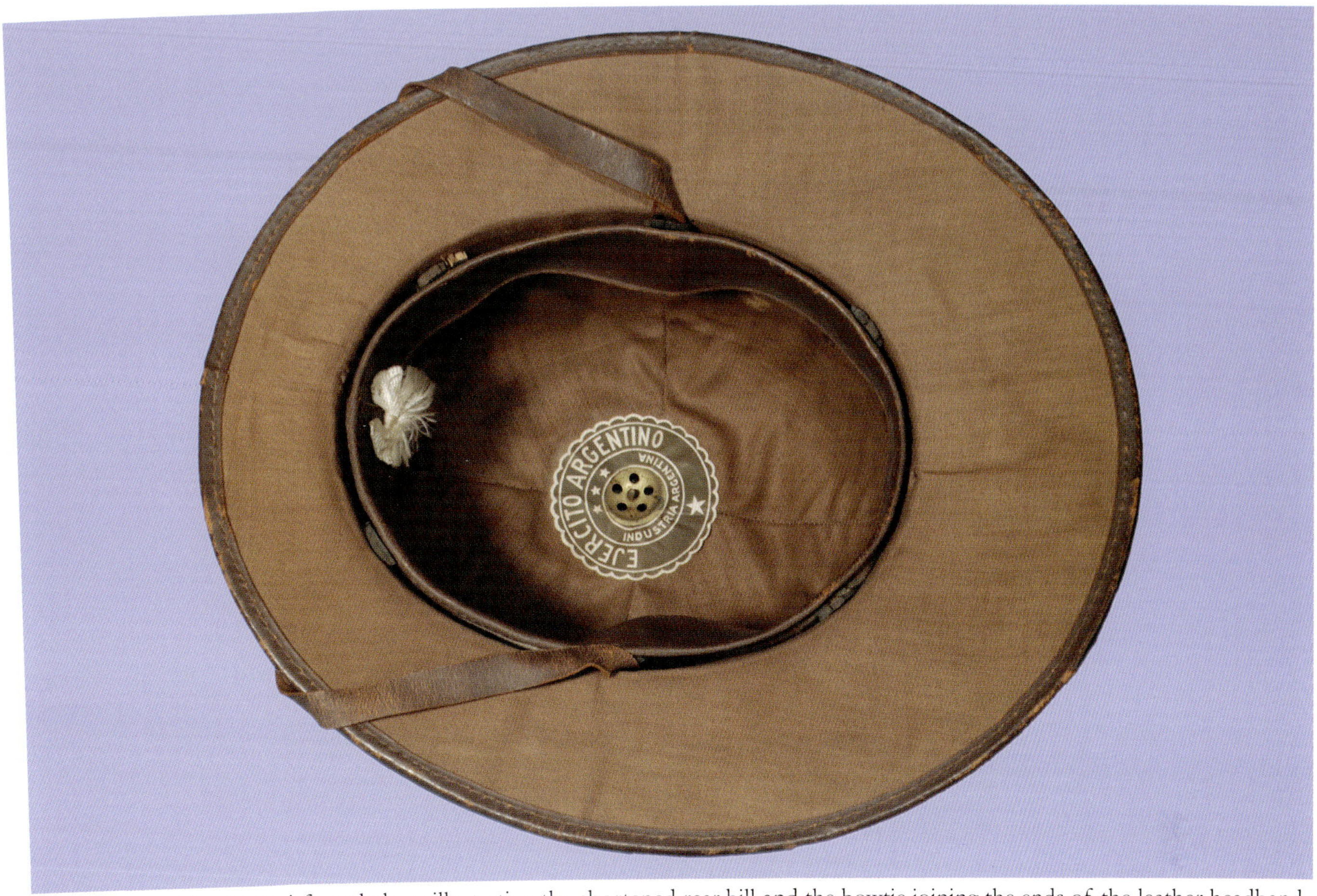

The interior of the 1927 airforce helmet illustrating the shortened rear bill and the bowtie joining the ends of the leather headband.

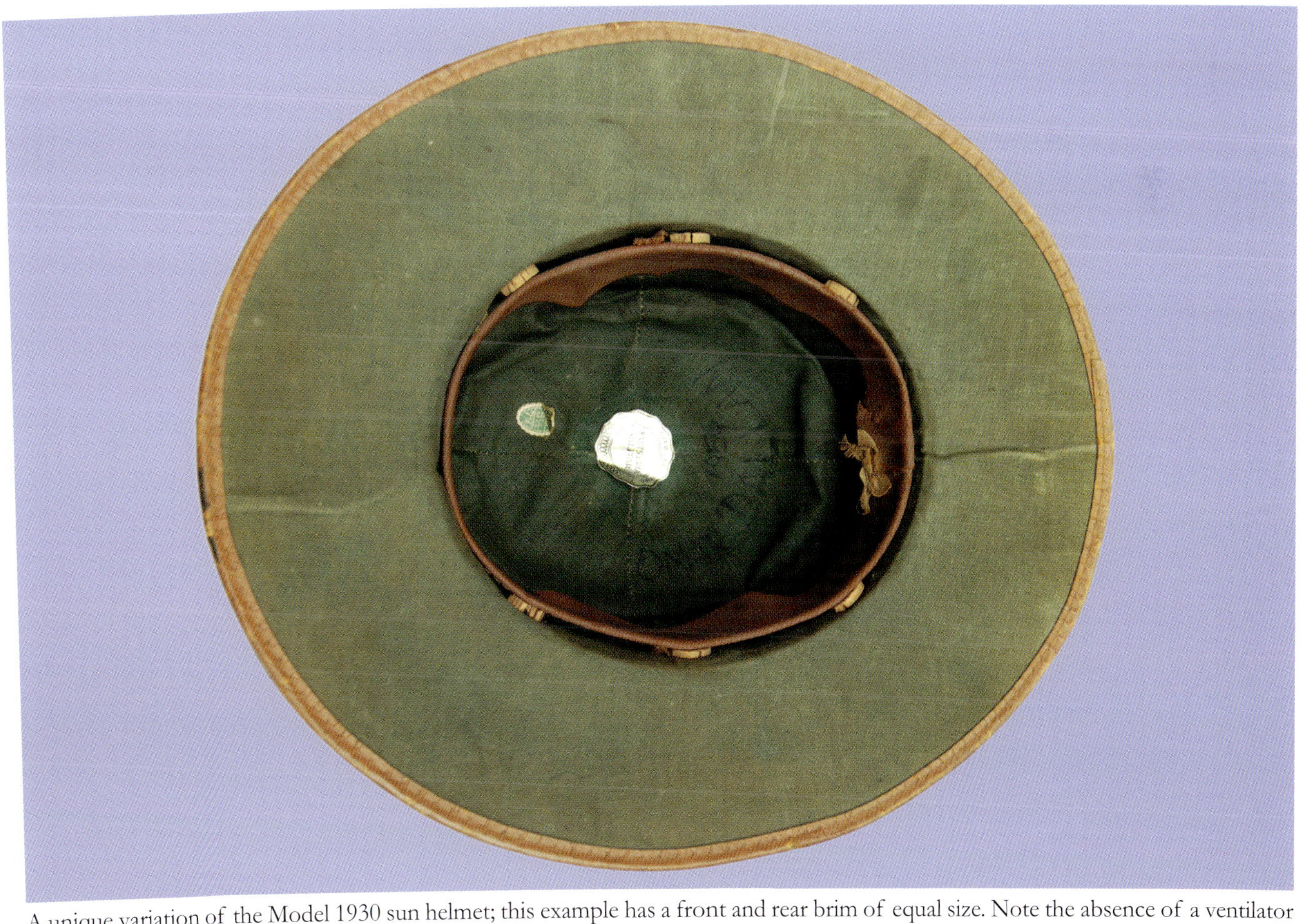

A unique variation of the Model 1930 sun helmet; this example has a front and rear brim of equal size. Note the absence of a ventilator at the top of the helmet which had only two small vent holes to the front instead.

Argentine troops on summer maneuvers in 1934. They are wearing the casco Italiano helmets.

A new pattern helmet of the "safari" style was introduced around 1927. It is sometimes called the "casco Italiano," after the similarly shaped Italian Model 1928 helmet. It is likely that this pattern was introduced prior to the Italian model in fact, but as Argentina looked to Italy in the early 1930s as a Fascist partner it is possible that similarities were seen and thus the nickname came about later.

The helmet had a rounded dome made of four panels, with a small vent and grommets in the center of each panel, as well as a standard ventilator at the top of the helmet. It was wrapped with puggaree around the helmet and featured a fairly wide bill around the base of the dome. This helmet is unique in that it has a longer bill or visor to the front than in the back – but because some examples are encountered with badges at the back, as well as photos showing soldiers wearing it both ways – it was never clear which was the true "proper" front or back of this helmet. To possibly provide "help" for the wearer the thin leather headband is tied with a cloth bowtie. Six or eight cork bumpers supported the leather band and further provided ventilation for the wearer.

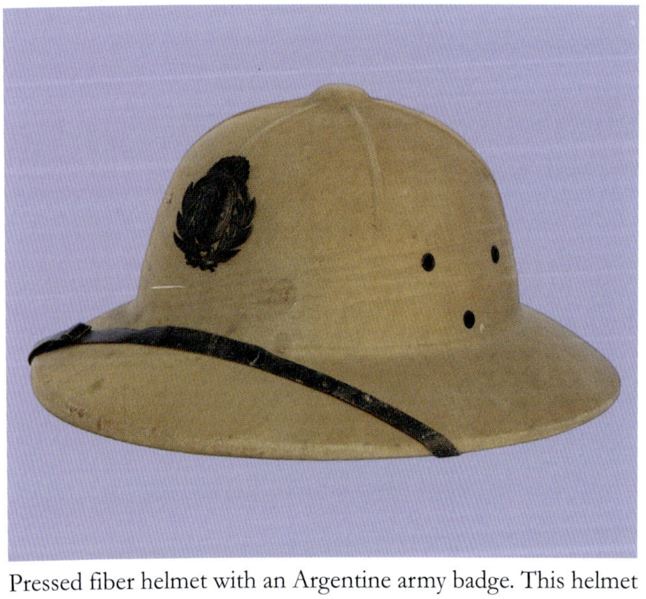

Pressed fiber helmet with an Argentine army badge. This helmet pattern was introduced in World War II and was likely based on the American pressed fiber helmet, and made domestically in Argentina. Note that the placement of the vent holes differs from the American made counterparts. (Courtesy of the collection of Eduardo Antonio Krapovickas).

These helmets were produced locally in Argentina, and were used with and without insignia on the front of the helmet. These included a smaller version of the aforementioned army badge that had been used on the Argentine spiked helmets. Additionally helmets were used with military academy badges, unit insignia and often with the national colors light blue and white (light blue–white–light blue) cockade. The "casco Italiano" remained in use until the 1960s when these helmets were widely retired from service.

It is worth noting that Argentina did export the "casco Italiano" pattern helmets to its neighbors, including Uruguay and Brazil. But the country also imported American pressed fiber helmets, which were used with cockades and badges. As Argentine officers and cadets often had to purchase their own headgear, and the regulations appear to have been lax on the subject of helmets, it seems likely that some soldiers simply looked for more affordable alternatives.

Brazil

Perhaps because of its Portuguese heritage, which had practically no relations with Germany, Brazil was one of the few South American nations that didn't follow Prussian style with spiked helmets. Following the First World War the nation did adopt a unique type of helmet that has been classified as a "sun helmet," and is included in this text – although technically it may not have been devoted to the purpose of providing protection from the sun at all.

This was a helmet made of pressed fiber, which was

A side profile of the Brazilian fiber helmet. These helmets were used during the Brazilian Revolution of the 1930s.

Modeled after the French Adrian Model 1915 and Model 1926 pattern helmets, the Brazilian Army produced a fiber helmet that offered no ballistic protection. These helmets are waterproof and would provide reasonable protection from the sun.

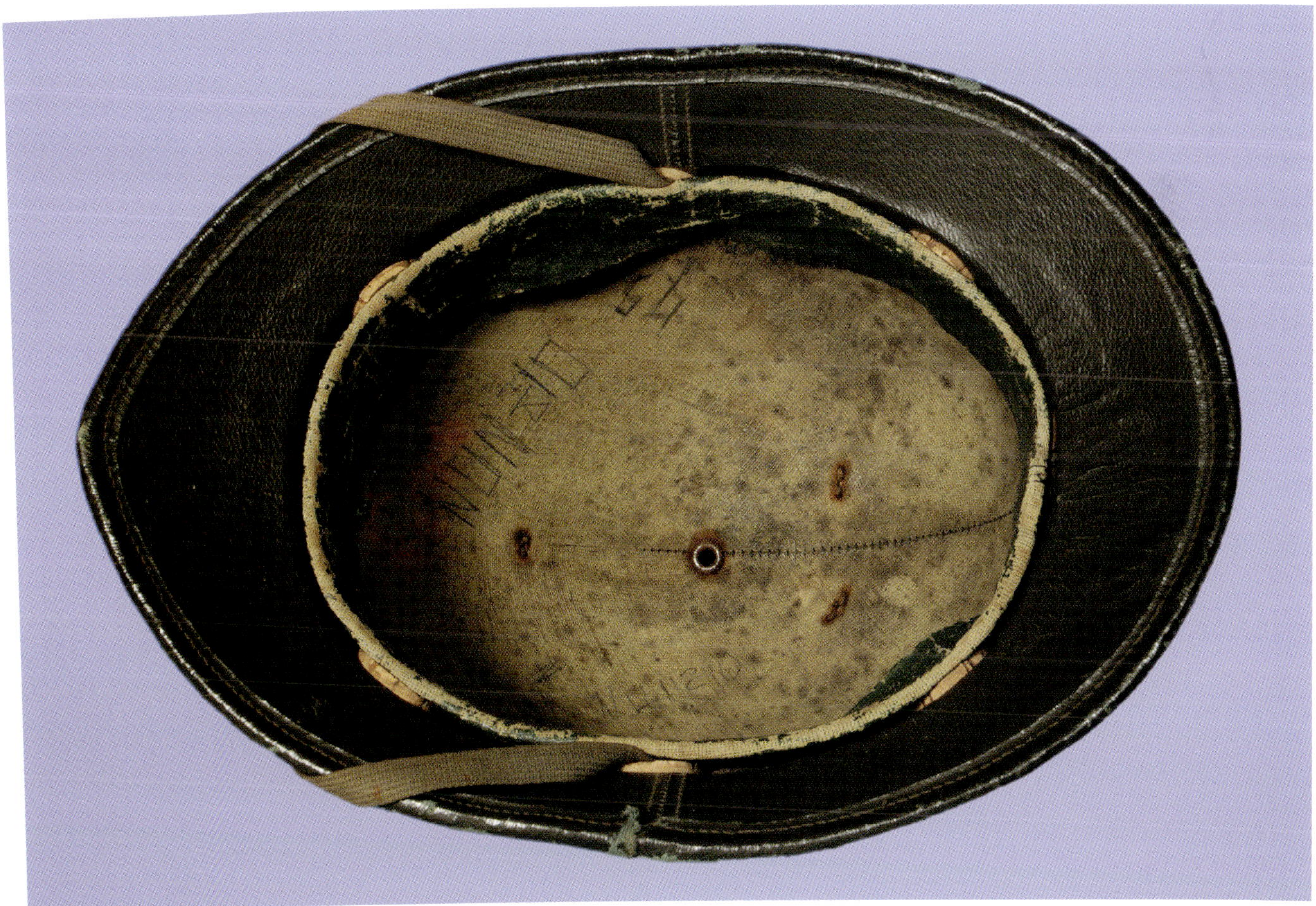

The liner of the Brazilian fiber helmet. While the exterior of the helmet is very similar in size and shape to the French Model 1915 steel helmet, the liner is completely different.

impregnated with a bonding agent, covered in a canvas cloth and painted to make it waterproof. The helmet was modeled after the French Model 1926 steel helmets, but as it wasn't steel it offered absolutely no ballistic protection. These were probably good for the jungles of Brazil, and while they lacked protection from even the smallest of small arms, they likely stood up to the elements quite well.

This helmet had the same profile as a French "Adrian" helmet, complete with a comb on the top, but with a more pointed front brim than the French steel pattern. The helmets were produced in a variety of shades of green, and featured a leather headband along with multiple cork spacers. A canvas chinstrap was mounted to one of the cork spacers.

And while the French-inspired helmet was likely the most common pattern, a number of variations exist, including fiber helmets that looked like a cross between the French M26 and the British MkI steel helmet.

On the front of these helmets was typically worn a regimental or arms of service device, and these were placed through a hole with grommet at the front. These fiber helmets were widely used in the Brazilian Revolution of 1930, and were eventually replaced by steel helmets.

Peru

Unlike many of its Latin neighbors Peru did not follow Prussian or German lines in regards to its military fashion. The nation modernized throughout the 1930s, and adopted a "safari" pattern sun helmet, which was used by officers and enlisted men as part of the summer uniform.

The helmets lacked any significant model number or pattern and are simply known as "sun helmets." As Peru followed French military patterns, including using the French Model 1926 helmet, it is believed that the Peruvian helmets were based on the French Model 1931 sun helmets, and the helmets have the same basic shape.

These helmets were made domestically in Peru, and were constructed of cork with six panels of a dark olive drab over the dome. There were two vent holes with grommets on each side of the helmet along with a faux ventilator cap at the top – in most examples the ventilator is mounted to the helmet but there is no vent hole to the dome present. A single band of cloth went around the base of the dome, but no puggaree was worn with these helmets. The headband was of the same thick rubberized leather that was used in the American pressed fiber helmets and these had a very rudimentary design.

The helmets were used during the Ecuadorian-Peruvian Border War of 1941-42, but were generally phased out by the 1950s.

The Peruvian sun helmet features a maker's label on the inside, with a Peruvian military label over the maker's label.

This Peruvian sun helmet is likely based on the French Model 1931 pattern sun helmet, and was introduced in the 1930s. It was used by the Peruvian Civil Guard Cavalry during the Ecuadorian-Peruvian war in 1940. This example has the army cockade and army officer insignia. It features a maker's label on the inside, with a Peruvian military label on top.

The label of the Peruvian sun helmet above.

Uruguay

The other nations of South America also used a variety of sun helmets on occasion, and this included the small nation of Uruguay. Throughout the 1930s and 1940s, Uruguayan military officers were required to buy their own uniforms, and often used Argentinean "casco Italiano" style helmets as well as American, and even the occasional British and French surplus helmets. During the Second World War Uruguay also imported pressed fiber sun helmets of the American design, although it is not clear whether these were made in the United States or Argentina. The addition of Uruguayan cockades and arms of service badges on the front is what distinguishes these helmets.

A pressed fiber helmet of American design. These were a popular private purchase helmet, as they were generally durable and more affordable than the canvas covered cork helmets. This example features the arms of service device for the artillery.

An undated photo of Uruguayan officers wearing a variety of sun helmets with an equal variety of cap insignia.

A private purchase helmet with the cockade and army insignia of the Uruguayan army. This sun helmet pattern is likely based on the French Model 1931, but was manufactured either domestically in Uruguay or possibly Peru some time prior to the outbreak of the Second World War.

Other South American Countries

Sun helmets were often not part of the typical uniform for many South American armies, but instead were brought into use during times of war. This is notable in the Chaco War (1932-1935), which was fought between Bolivia and Paraguay for the desert region known as the Gran Chaco.

Period photographs suggest that some officers on both sides used sun helmets, likely of European and South American origins. Due to the placement of each nation to its neighbors – and the fact that both are landlocked nations – supplies came from the closer friendly countries. Thus it is not surprising to find Argentinean sun helmets with Paraguayan badges.

A Paraguayan officer's helmet with insignia attached to the front. This example appears to have been painted at some point, possibly to make it waterproof or just to match the tan colored uniforms used by Paraguay.

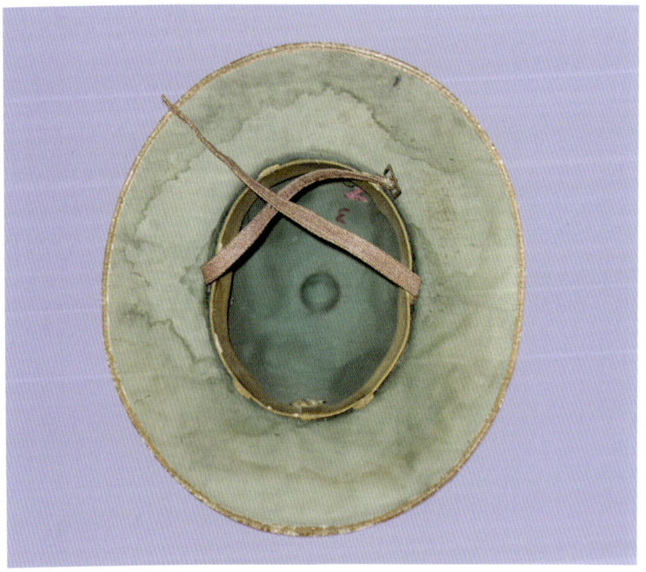

The interior shows that this is likely an Argentinean made helmet for private purchase, and has the basic shape and design of the French Model 1931. While crudely made these would have been more affordable to those officers looking to wear a more distinguished type of headgear than the soft bush hats worn by the bulk of the army in the Chaco War.

The same basic policies also held true for numerous Latin American nations, and thus a variety of helmets can be encountered. It is stressed that the buyer use their own judgment, especially given that cockades from South American countries are available in abundance!

A group of Paraguayan soldiers, most likely officers, stand at Fort Muñoz, former headquarters of Bolivian General Kundt, which was captured on December 19, 1933.

CHAPTER 9

VIETNAMESE - RUBBERIZED CARDBOARD HELMETS OF SOUTH EAST ASIA

The last major conflict to see the widespread use of sun helmets by a combatant force was of course in South East Asia during the Vietnam War – also known as the Second Indochina War. Today the distinctive sun helmets worn by the People's Army of North Vietnam remain one of the war's strong iconic symbols, possibly only overshadowed by the robust AK-47. The sun helmets of Vietnam fall into several categories including those actually used before 1975 as well as the various versions that were used in the following years.

During the First Indochina War with France, the Vietminh used surplus Japanese sun helmets but few survived to see service more than a decade later in the conflict with the Americans. Additionally, the earliest examples of sun helmets used by the Vietminh were made locally of bamboo and other fibers that were woven together and cloth covered. These were often used with netting and scrim to provide some camouflage. A basic Vietminh star was often pinned to the front of the cloth, but it is believed that only officers wore the badges or pins. This pin was usually a metal round disc with red enamel paint and a strike out gold star. This design was used from the late 1940s up until about 1958 when it was replaced with a modified design that utilized a smaller star that was less rounded. The star was framed with two wreaths and sat above an industrial-styled cog.

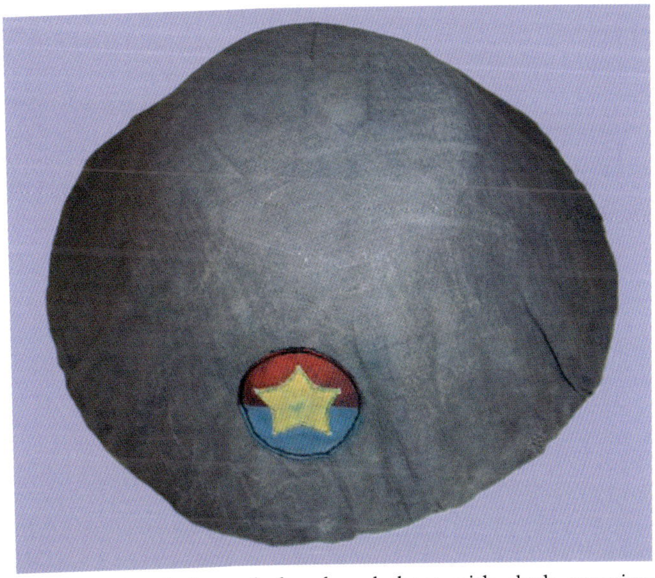

An example of the early bamboo helmet with cloth covering (Private collection).

The interior of a bamboo helmet with cloth covering. This design was used from the late 1940s to about 1958 (Private collection).

Second Indochina War Helmets

After the defeat of the French in 1954 and the division of Vietnam into two distinct nations the forces of North Vietnam – known as the People's Army of Vietnam – adopted a blended mix of Soviet and Chinese styled uniforms and equipment. The badges were further refined and came in three basic ranges, which were reflective of rank from general officer, field grade officer and enlisted men. Beginning in 1958 branches at all levels wore the badges, which were denoted by color and device, with an anchor on a dark green background for naval Coastal Defense Units (PAVN Navy) and wings on a blue

This wartime Vietnamese Pith Helmet features the VC badge. Used by the North Vietnamese and National Liberation Front forces during the Vietnamese War, this basic helmet design remains in service today. Made of a synthetic material with a cotton cloth layer bonded to the helmet the liner is rubberized cloth. This wartime example features a leather chin strap and has a steep dome and rim.

This late wartime example has a more gradual slope to the helmet bowl and rim. It features the traditional badge of the PAVN (People's Army of Vietnam), also known as the NVA.

background for airforce. During the Second Indochina War cap badges were ordered removed, no doubt in an attempt to maintain that the troops were volunteers fighting for the cause of the Vietcong, rather than a true invasion by Northern forces.

While the Vietcong and even the People's Army of Vietnam – sometimes referred to as the NVA (North Vietnamese Army) – relied heavily on outside support in the way of equipment, weapons and most importantly ammunition, the sun helmets were actually domestically produced. It was one of only a few items used in the military struggle that was made in the country. Introduced around 1964 the first helmets were made of a synthetic

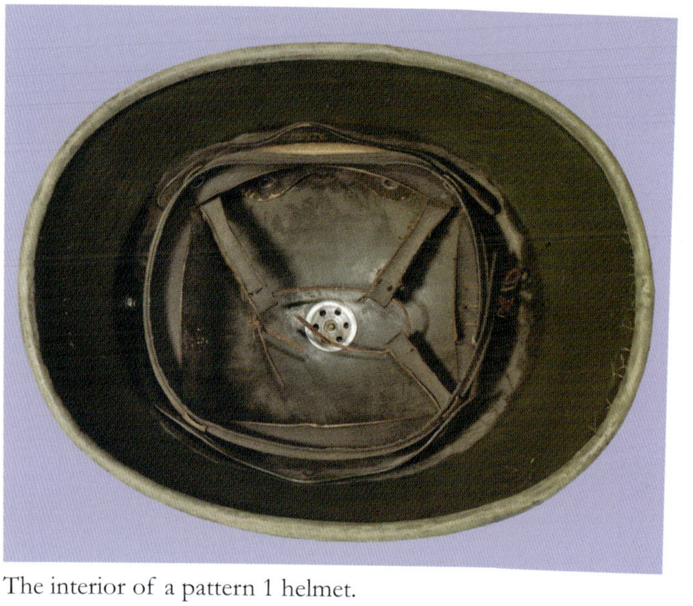

The interior of a pattern 1 helmet.

material called Phenolic, which is a type of cardboard that is impregnated with resin under pressure. The helmet was covered with a cloth layer that is permanently bonded to the helmet. Cloth of various colors was used and hence there was no one standard; coupled with exposure to the sun and elements, today helmets in a range of colors are encountered. During the war the helmets were produced in a variety of shades, which included green, tan, brown and in the case of the PAVN Navy in white.

The helmet liners were rubberized cloth, which was mounted to the helmet with soft metal rivets. These helmets typically used a leather chinstrap, although a plastic chinstrap was introduced sometime after 1973 and may be found on some late period helmets. This style helmet was actually produced under an East German license but it should be stressed that this does not imply that the design was similar in any way to the World War II era German sun helmets. A variety of shapes were produced throughout the war time era, with the earlier helmets having a distinct dome shape that eventually was

The interior showing the rubberized headband with a leather chinstrap.

An infamous photo of actress and activist "Hanoi" Jane Fonda with the North Vietnamese Army in 1969. While Ms. Fonda is wearing a steel helmet, likely of Soviet manufacture, this photo offers a good look at the sun helmet used by the PAV.

The distinctive Cong An badge.

An example of the PAVN (Navy) helmet badge.

replaced with more of a "safari" sun helmet shape. The basic helmet remained in service following the war, with minor improvements to the rubberized liner and plastic chinstrap.

The cap badges were standardized for all branches of the People's Army of Vietnam in 1982, and the army's red cap badge became the new standard. The exception was certain security police units, which were re-organized under the Ministry of the Interior. These units continued to wear the distinctive Cong An cap badge.

Post War Helmets

In 1983 a new version of the sun helmet was introduced and this type remains in service with the Vietnamese

Army of the Socialist Republic today. Usually seen in a bright green color, with a diamond shaped label this helmet is typically encountered online as a wartime helmet. This label was not in widespread use during the

North Vietnamese soldiers on the march.

This post-war sun helmet is that used by the Vietnamese Army of the Socialist Republic since 1983. It is a modernized version of the wartime helmet. It features a plastic liner and chinstrap. This helmet features the Vietnamese Cong An (Military Police) helmet badge.

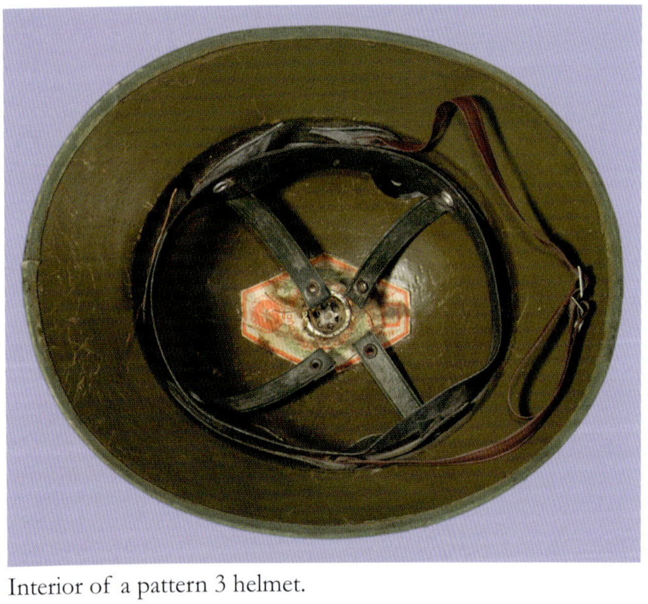

Interior of a pattern 3 helmet.

wartime era and this is just one sign that such helmets might in fact be of post-war manufacture. Likewise a lighter version, with even darker green cloth, is still being manufactured today for sale to tourists.

It is also worth noting that there were actually "period" wartime copies of the helmets made in Taiwan for use by Allied reconnaissance teams, but these typically lack the waterproof liners. The early models featured leather liners while plastic was common in the later helmets. Very few of these examples are believed to have survived the war however.

Chapter 10

Miscellaneous Sun Helmets

In addition to those nations already covered by previous chapters it is worth noting that many other armed forces also used sun helmets. It is impossible to cover every single example. Suffice it to add that sun helmets were used by many nations for a variety of reasons, including for naval personnel at times when a vessel would make port of calls to tropical regions, for training in hot weather climates and as part of summer uniforms. Additionally, at various times many nations allowed sun helmets to be part of a walking out uniform for officers, who often bought their own private purchase items. This makes covering every example almost entirely impossible.

For the purposes of including as much information on the subject of military sun helmets, various examples – including regionally produced helmets – have been included. It is also worth noting that during the Second World War a seemingly limitless number of variations were produced and used by the combatant (and even non-combatant) nations. Additionally, many police, border guards and paramilitary groups have used sun helmets over the past 100 years, and again, it is simply not within the scope of this book to include every example.

Ceremonial Helmets of the 19th Century

Throughout the twilight of the 19th century many smaller nations adopted sun helmets as part of the uniform for ceremonial, royal and imperial guard units. Often times these units were meant to evoke the power of these nations, and to give a "Europeanized" appearance. As a result of the small size of these formations few of these helmets survive today, but records on those used have shed some interesting insight.

The small island nation of Hawaii was ruled by a king who had absolute power, and in the 1880s King Kalakaua I was just such a ruler. He modernized his small army in the style of the major European powers, and this included the King's Guard. These soldiers wore white cork sun helmets, likely of British or American manufacture, which were fitted with a brass spike and chinchain that passed around the body of the helmet and attached to a hook just behind the spike. At the front was a badge of the king's monogram "Kalakaua I" with a European style crown above.

Half way around the world at the same time another small nation was modernizing its equally small Imperial Guard. This was the Guard that served as part of the puppet army of Queen Ranavalona III of Madagascar, and this unit wore white sun helmets that were likely of French manufacture. These featured a brass spike – something not utilized ever by the French – along with an Imperial crown at the front as well as an eagle in a wreath. The only known surviving example is today in the Musee de l'Armee in Paris. These helmets were worn proudly until relations between the Queen and France disintegrated. This was followed by the French invasion which transformed Madagascar into a formal French colony.

Imperial Guards helmet of Madagascar. This is the only known survivor and is housed in the Musee de l'Armee, Paris.

A Dutch East Indies made helmet. This was one manufactured in Batavia (1619-1942), the present day Jakarta.

Interior label of the Dutch East Indies helmet above.

East Indies and Asian Helmets From the World War II Era

Prior to the outbreak of World War II there was a renewed sense of the imperialistic past in the Far East, and while it is apparent, in retrospect, that the Allied powers actually prepared inadequately for the coming conflict, the art of sun helmets expanded dramatically through the region.

A sun helmet industry, which had been well established in British India, spread to the Dutch East Indies, Malaya, Singapore and Hong Kong. The numerous surviving examples indicate that private purchase helmets were produced in apparently large quantities. A number of materials were used throughout the late 1930s including cork and interestingly enough sola pith, the latter making a comeback as war approached. Surviving examples also show that newspapers and other materials were used to create the bodies of the helmets, and often times the production methods seem almost crude, with many helmets lacking a puggaree – although an equal number of helmets are of very high quality with puggaree and even silk linings.

Many commercially made civilian helmets were

Star Brand label of the Hong Kong made sun helmet.

A Hong Kong made helmet featuring the Star Brand label to the interior.

A World War II period private purchase helmet from Malacca. The helmet may have been based on the American pressed fiber helmets as it features a faux puggaree that is then wrapped with a real single band of puggaree cloth.

The inside of theMalaccan private purchase helmet, this example features a notably impressive maker label that includes a B-29 bomber!

pressed into military service during the Second World War, often used by Allied officers in the jungles of Burma and New Guinea, as well as by Japanese officers throughout their vast conquered territories. This has presented a problem for collectors however, as it is very easy for a faker to affix a simple Japanese star, American badge or British shoulder flash and claim to have a field-service helmet. Collectors are advised to do their homework, such as looking at labels to determine where the helmet was produced and if possible when.

Ethiopian Sun Helmet

Unlike the various "liberation" armies of Vietnam, most post-independence African nations never widely adopted sun helmets with the exception of Ethiopia. Prior to the late 19th century "scramble for Africa" sun helmets weren't widely used by the various powers in Africa either, and the first sun helmets likely arrived with explorers such as Livingston and Stanley, and of course by the British soldiers in red coats – followed by those of the other nations already included in this text.

Ethiopia had already been a nation exposed to men in sun helmets by the time Italy looked to expand its empire in 1896. The small Abyssinian Empire (as Ethiopia had been known) had faced a small invasion by the British over a hostage situation in 1867, and was a pawn of the major powers by the 1890s, receiving arms from France as well as other military equipment. A brief war with Italy ended that European nation's hopes of expanding its East African empire at the battle of Adowa. Following this victory, Ethiopia continued to modernize and by the outbreak of the First World War was the only nation in Africa (besides tiny Liberia) to retain its independence.

In the 1920s and 1930s Abyssinia, with the aid of

Emporer Haile Selassie wearing a British Wolseley while visiting his people in Ethiopia prior to the 1974 revolution, which saw him overthrown.

An example of the Ethiopian Polo Style helmet.

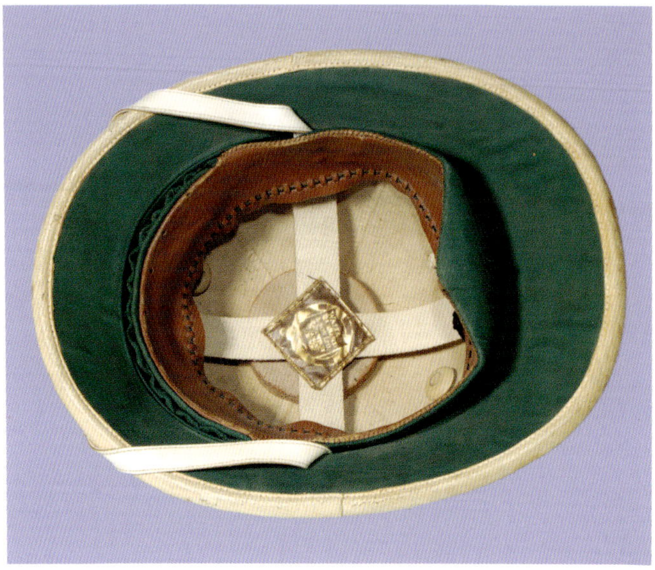

Interior of the Ethiopian Polo Style helmet.

Swiss, Belgian and Swedish volunteers modernized the army. This included European style uniforms and notably sun helmets. These were based on the British Wolseley pattern – and likely were surplus Wolseleys as well as Belgian-made versions. These were typically worn without insignia or devices by much of the Abyssinian army, while the Zabagna gendarmes – among the elite forces – did wear cockades on the front, likely in the national colors, which were red, yellow and green.

Despite all efforts to hold out, as well as with support from various Western volunteers, the Italian army defeated the Ethiopians in May of 1936, sending Haile Selassie into exile. He would return with British help to liberate the nation following Italy's declaration of war on Great Britain in 1940. While the bulk of the invasion force was British, Selassie led remnants of the Ethiopian army in exile, and these men were outfitted with British equipment.

This included sun helmets, and interestingly the Ethiopians were outfitted with British "polo" style sun helmets featuring a narrow brim around the dome of the helmet. These were originally covered with a white cloth, made up of six panels around the dome, and this was likely stained to give it a more khaki-like color. The helmet did not have the usual top ventilator but instead had four vent holes with thick rubber grommets. Around the dome of the helmet was a puggaree and, following the British style, a flash of the national colors of Abyssinia was worn on the left side.

Following the war Haile Selassie was restored as Emperor, and reigned until the nation was gripped by revolution in 1974. During the 1960s however, the nation once again attempted to modernize the military and this included the reintroduction of military sun helmets. This time the nation adopted a helmet pattern that closely resembles the British Wolseley. These helmets were originally imported from the U.K. as well as Belgian Congo, but were later domestically produced.

The helmet featured the basic outline of the Wolseley, with six panels around the dome, as well as a top

An Ethiopian Cavalry helmet post WWII.

An Ethiopian Infantry helmet post WWII.

ventilator cap. A puggaree of four folds circled the helmet just about the brim. An arms of service badge was worn on the front of the helmet, and these included the Imperial crown at the top and the Star of David at the bottom (indicating that the Imperial family could trace its origin back to King Solomon and the Queen of Sheba). Helmets for the Imperial Guard or Royal/Imperial regiments also were adorned with the Ethiopian lion (the Lion of Judah, which is meant to represent Emperor Haile Selassie). The guards' helmets were often also adorned with a spike.

Interestingly the Ethiopians wore a tri-color cockade on the right side of the helmet, which included traditional colors of green, yellow and red. This was in contrast to the flashes being worn on the left side of the helmet. The cockades were supported with a thin piece of metal, and sometimes this metal was worn inside the dome of the helmet – likely to give support – but it is also encountered

A Polo pattern South African helmet used extensively in the North African campaign of WWII.

The Ethiopian cockade of green/yellow/red.

on the outside of the helmet, including in period photos. The purpose for wearing the shiny rectangular piece of metal has never been fully established. Following the 1975 revolution the helmets were no longer used, likely being seen as being a tie to the Imperial past.

South Africa, Rhodesia and Zimbabwe

While sun helmets were seldom used by the post-colonial powers in Africa, there were exceptions. The most notable was South Africa, and while the former British colony was part of the British Commonwealth at different periods, the nation's helmets are included in this section. The same holds true of Rhodesia, which later became Zambia (Northern Rhodesia) and Zimbabwe (Rhodesia).

The semi-independent Boer Republics had adopted a variety of European influenced equipment, including sun helmets for some artillery units. However few examples managed to survive the Second Anglo-Boer War (1898-1901). Following the war, the Union of South Africa was created in 1910 from the Cape and Natal colonies, as well as the former Boer Republics of Orange Free State and Transvaal. During the First World War the Union of South Africa fought alongside the Allies, and this included significant actions in German East Africa. During this time South African units wore a variety of British Wolseley and Cawnpore helmets, usually with South African flashes.

In the 1930s the British and Commonwealth forces in southern Africa used a number of sun helmets. These included locally produced "safari" style helmets, as well as surplus Wolseley and various "India" pattern helmets. The police and militia units wore these helmets with a range of flashes and puggarees.

It was during the Second World War that one of the rarest patterns of sun helmets appeared. It was a small brimmed sun helmet that has often been referred to as a "polo hat" because of its appearance. But whether these style helmets were ever actually used for polo is debatable. The helmet featured a hard cork shell that was covered with six panels of khaki colored cloth. The dome of the helmet was wrapped tightly with a puggaree of the same colored cloth, and many helmets feature a flash on

The interior of a Polo helmet showing the maker's stamp on the sweatband of Failsworth Hats Ltd.

the left side. At each side of the helmet above the flash were four ventilation holes with grommets, and notably no ventilator cap at the very top of the helmet. The inside of the helmet is covered in a green felt and the liner

A Rhodesian sun helmet with an unidentified flash.

consisted of a soft leather sweatband. Hooks for the leather chinstrap were attached to each side of the helmet dome.

These helmets were produced in large numbers by the Failsworth Hats Ltd. company and by the Vero Everitt Ltd. company, and based on surviving examples the main years of production were 1941 and 1942. These helmets were widely used by the South African forces fighting

Interior of the Zimbabwean helmet showing the maker's label.

Based on the post-war sun helmets from South Africa, this sun helmet was produced by Pith Helmet Industries in Bulawayo, Zimbabwe. It features the badge of the Zimbabwean Border Guards.

with the British 8th Army in North Africa, and remained in service with the post-war armies of South Africa and Rhodesia at least until the 1950s. It is also worth noting that following South Africa's withdrawal from the Commonwealth the nation was somewhat isolated from the world, but continued to develop military equipment. South African Pith Helmet Industry Ltd. (SAPHI), founded in 1940, developed the modern ballistic helmets of South Africa, which were introduced in the late 1980s.

And while Rhodesia became Zimbabwe in 1980 this nation continued to use sun helmets in some capacity to this day. While not a significant part of the modern Zimbabwe uniform, some border guard units do wear sun helmets. Pith Helmet Industries of Bulawayo, an offshoot of the South African Pith Helmet Industries, produced these helmets throughout the 1990s. These "safari" styled

A North Rhodesian police helmet with puggaree piping.

A helmet of the Royal Guards Brigade of Thailand.

helmets featured cork bodies covered with six panels of olive drab cloth around the dome, and oval brim. Around the helmet was wrapped a puggaree, with vent holes just above this. At the top was a faux ventilator cap, and some examples featured a crudely made arm of service badge.

Thailand

Thailand's Royal Guards Brigade is made up of soldiers hand-picked from the nation's military foot battalions. It boasts an ancient tradition that goes back some 600 years, to when Siam's elite archers successfully captured the city of Angkor in Cambodia and took nearly 90,000 prisoners in the process. The unit wears a variety of uniforms including one of a traditional Siamese costume. However, most visitors would likely see the white palace guard uniform, which was complete with a white summer sun helmet.

The white helmets have been used for nearly 100 years and feature a basic "safari" pattern shape, likely being influenced by the similar helmets of Anglo-Indian origin. For all this time the helmets were fitted with a tall spike at the top, along with the royal crest on the front. At the sides of the helmets were rosettes, which appear to have imitated those of British pattern Home Service Helmet. Over the years the helmets were made of cork and sola pith, covered with a white cloth, but plastic helmets have also been used during the local "rainy season."

In addition to the ceremonial helmets, Thai forces have used sun helmets in the field. After World War II the

Based on the French pattern sun helmets used in French Indochina, this Thai Border Patrol sun helmet features the Thai royal crown.

Thai military relied on much surplus Japanese equipment, and reissued the Type 90 steel helmet with a Thai badge on the front, but the armed forces also used Japanese produced tropical helmets. More recently the nation has used its own domestically produced sun helmets for border patrol units. These were lightweight cork helmets of six panel construction, covered with a dark khaki cloth, and featured the arms of service badge at the front.

Greece

While lacking any "overseas" colonies – at least in the modern era – the Southern European nation of Greece isn't short on sun. Thus, beginning in the 1930s the Greek military began to use sun helmets as part of the summer uniform. While not a private purchase item, these helmets were generally issued to officers, as well as guards and others who had to spend a great deal of time under the warm Eastern Mediterranean sun.

The helmets were locally produced, and were a copy of the British Wolseley pattern sun helmet. These featured a flared out brim but with a pointed front visor. The helmets were made of cork, and featured a six-panel construction. Instead of a puggaree a thick band of canvas was wrapped around the base of the dome. On each side of the center of the dome were three vent holes with grommets, while the top of the helmet had a ventilator cap to provide adequate airflow for the wearer. The outside of the helmet was covered with white canvas cloth, while the inside was a dark green cloth and leather sweat band. The helmets were typically painted with a thick white blanco, which was applied whenever the helmet showed excessive dirt. A variety of military badges was worn at the front of the helmets, and these appear to have been very popular with the Greek Air Force. These helmets remained in service until the 1970s.

Additionally, a large number of helmets were captured by the Germans during World War II and these were used

The Greek army used a sun helmet based on the British Wolseley pattern. These were introduced in the 1930s and remained in use until the 1970s. This example features the insignia of the late 1960s era Greek Air Force.

Front view of the Greek Air Force sun helmet.

by German forces in the Balkans, as well as Kriegsmarine personnel, who were especially fond of this particular pattern.

Many collectible books offer information on the collecting, care and preservation. So it goes without saying that collectors should use common sense when it comes to caring for and preserving the items in their collections. Sun helmets are much more fragile than steel helmets obviously. And while these may have spent countless days or weeks under the heat of the sun for the sake of preservation sun helmets are best displayed in a cool, dry room and most importantly out of direct sun light.

It is this author's opinion, which is shared by many colleagues, that sun helmets should not be handled with bare hands. Rather than fading from light, sun helmets tend to darken from repeated handling by dirty or greasy hands. Likewise, dust is the often much more a concern than it would be for other helmets.

While sun helmets may never obtain the collector appeal of the steel combat helmets, these helmets deserve as much respect and care.

While there is obviously much more information yet to be revealed about this headgear, it is the author's hope that this book has revealed some new facts, and shed some light on a subject that has previously only been covered as a section in other helmet or militaria-related books.

Abbott, Peter, Armies of East Africa 1914 –18, Osprey Publishing, Oxford, 2002

Abbott, Peter, Colonial Armies Africa 1850 to 1918, Foundry Books, Nottingham, 2006

Attard, Robert, Collecting Military Headgear: A Guide to 5000 Years of Helmet History, Schiffer Publishing, Ltd., Atglen, Pennsylvania, 2004.

Barthorp, M, British Infantry Uniforms Since 1660, Blandford Press, Dorset, 1982.

Barthorp, M., The British Troops in the Indian Mutiny, Osprey Publishing, Oxford, 1994.

Boulton, Dr. James J., Head-Dress of the Canadian Mounted Police: 1873-2000, Bunker to Bunker Publishing, Calgary, 2000.

Branco, Pedro Soares, Coberturas Militares Portuguesas: 1740-1918, Edições Inapa, 2000.

Carman , W.Y., Indian Army Uniforms, Morgan-Grampian, London, 1969.

Chappell, Gordon, Summer Helmets of the U.S. Army 1875-1910, Wyoming State Museum, Cheyenne, 1976.

Chappell, Gordon, Brass Spikes and Horsetail Plumes: A History of U.S. Army Dress Helmets 1872-1904, Thomas Publications, Gettysburg, Pennsylvania, 1997.

Chappell, Mike, The British Soldier in the 20th Century: Field Service Head Dress 1902 to the Present Day, Wessex Military Publishing, 1987.

D'ami, Signor Rinaldo, World Uniforms in Colour Vol. 1, Patrick Stephens Limited, London, 1968

D'ami, Signor Rinaldo, World Uniforms in Colour Vol. 2, Patrick Stephens Limited, London, 1969

Davis, Brian, British Army Uniforms and Insignia of World War Two, Book Club Associates, London, 1983.

Dress Regulations for the Army, 1900, David & Charles (Publishers) Limited, Devon, 1970.

Festberg A. & Videon B., Uniforms of the Australian Colonies,Brown, Prior & Anderson, Victoria, 1971.

Figueroa, J.R., Tropical Headgear of the German Wehrmacht in W.W. II Vol. 2, Figueroa Creations, Los Angeles, 1996

Grebert, R., Military Badges of the Australian colinies 1800-1903, NSW Military Historical Society, Sydney, 2000.

Haythornthwaite, Philip, Uniforms of the Civil War 1861-65, Macmillan Publishing Co., Inc., New York 1975.

Heathcote, T.A., The Indian Army, Wren Publishing, Melborne, 1974.

Hennequin, Robert, French Military Helmets of the XX Century: History and Identification (1945-1970)/Casques Militaires Français Du XXe Siécle: Histoire et Identification (1945-1970) Tome 2, Hénricy 2006.

Howell, Edgar M., United States Srmy Headgear 1855-1902: Catalog of the United States Army Uniforms in the Collecitons of the Smithsonian Institution, Vol 2, Smithsonian Institution Press, Washington, D.C., 1975.

Johansson, Eric, Pickelhauben (Spiked Helmets), H.S.M. Publications, Independence, Missouri, 1982.

Jowett, Philip S., The Italian Army 1940-45 (2), Osprey Publishing, Oxford, 2001.

Jowett, Philip S., The Japanese Army 1932-45 (1), Osprey Publishing, Oxford, 2002.

Jowett, Philip S., The Japanese Army 1931-45 (2), Osprey Publishing, Oxford, 2002.

Kasal, Mark, and Moore, Don, A Guide Book to U.S. Army Dress Helmets 1872 – 1904, North Cape Publication, Inc., Tustin, California. 2000

Lang, Otto Peter, Deutsche Stahl-, Tropen- und Lederhelme: 1916-1946, Weishaupt Verlag, 2005.

Langellier, John P., Hats Off: Head Dress of the U.S. Army: 1872-1912, Schiffer Publishing, Ltd., Atglen, Pennsylvania, 1999.

Marzetti, Paolo, Elmetti/Helmets, Parma, 2003.

Mollo, John, The Formation Sign, The Military Heraldry Society, FS158, April 1990.

Nicolle, David, The Italian Invasion of Abyssinia 1935-36, Osprey Publishing, Oxford, 1997.

Nila, Gary & Rolfe, Robert A, Japanese Special Landing Forces Uniforms and Equipment 1932-45, Osprey Publishing, Oxford, 2006.

Peers, Chris, Armies of the Nineteenth Century: Africa East Africa, Foundry Books, Nottingham, 2003

Quesada, Alejandro de, The Spanish-American War and Philippine Insurrection 1898-1902, Osprey Publishing, Oxford, 2007.

Rankin, H., Military Headdress, Arms & Armour Press, London, 1976.

Rest, Stefan (Editor), The Emperor's Coat In the First World War, Verlag Militaria, Vienna, 2002.

Schiers, Ulrich, Tropenhelme der kaiserlichen Marine, der Ostasiastichen Truppen und der Schutztruppen, Deutsche Gesellschaft für Heereskunde e.V., Mechenheim, 2007.

Tulkoff, Alec, Grunt Gear, James Bender Publishing, San Jose, 2003.

Wedd, M., Australian Military Uniforms 1800-1982, Kangaroo Press, Sydney, 1982.

Wilkinson-Latham, C., The Indian Mutiny, Osprey Publishing, , Oxford, 1977.

Wilkinson-Latham, Robert and Christopher, Infantry Uniforms Book Two 1855-1939, Blandford Press, London, 1970

Wilkinson-Latham, Robert and Christopher, Home Service Helmet: 1878-1914 With Regimental Plates, The Broadway Press Ltd., Star Products .